RESTRICTED AND GENERAL DIMENSIONAL ANALYSIS

RESTRICTED AND GENERAL DIMENSIONAL ANALYSIS

TREATMENT OF EXPERIMENTAL DATA

Constantin I. Staicu

ABACUS PRESS
Tunbridge Wells, Kent

Abacus House, Speldhurst Road, Tunbridge Wells, Kent, England

Revised, up-dated English edition of the Romanian work

ANALIZA DIMENSIONALĂ GENERALĂ
first published by EDITURA TEHNICĂ, Bucharest

Translated by Dinu Manoil and Alexandru Cavaleru
Translation edited by Patricia Monahan

BL British Library Cataloguing in Publication Data

Staicu, C I
Restricted and general dimensional analysis.
1. Dimensional analysis
I. Title II. Analiza dimensională restrînsă și generală
530'.8 QC20.7.D55
ISBN 0-85626-300-1

Printed in Romania

Knowledge is power.

Bacon

Preface

Dimensional analysis has been the concern of many mathematicians throughout the ages : the Ancient Greeks, Ptolemy and Pappus from Alexandria and later, Galilei, Descartes, Newton and Leibniz. The original work by Euler in this field was developed by Lagrange and Laplace, who expounded the principles of dimensional analysis. It is now agreed that the basis of dimensional analysis was laid by Fourier, in the first quarter of the nineteenth century ; indeed, he was the first to provide a definition for the dimensional homogeneity of physical relations and applied the geometry concept of "dimension" to physical quantities. In 1848, Bertrand tried various combinations of physical quantities in his attempt to obtain non-dimensional products. Important developments in this field were made after 1870, when Maxwell described the dimension as a "dimensional equation", the meaning now used ; furthermore, Bertrand showed that physical relations could be derived on the basis of the dimensional homogeneity condition, provided the physical quantities involved in the said relations are known ; Vaschy stated the Π theorem while Rayleigh expressed a dependent physical quantity through the product of powers of the independent quantities which describe it and applied the non-dimensional product to fluid mechanics. Early in this century, Buckingham gave a more accurate formulation for the Π theorem ; Bridgman made some specific contributions to the number of non-dimensional products of powers and Einstein applied dimensional analysis to estimate the oscillating frequency of atoms in a solid body. Rayleigh is considered to be the founder of dimensional analysis and Bridgman the greatest authority in the field. In the second half of the XXth century, Huntley described the method of components of fundamental dimensions and Langhaar has demonstrated the theorem of (restricted) dimensional analysis, while Charcosset first gave a critical review of the older demonstration of the dimensional analysis fundamental theorem and then suggested a new one.

Bridgman, Esnault-Pelterie, Martinot-Lagarde, Langhaar, Duncan, Palacios, Comolet, Huntley and Douglas published, after 1920, works devoted to dimensional analysis.

All these books deal with the old dimensional analysis; here we have used the term *restricted dimensional analysis*, because this kind of analysis helps to find the values of quantity exponents of physical relations in that particular case when only one non-dimensional product of powers is involved, that is when only a small number of relations of measurable variables are correlated.

The author uses the term *general dimensional analysis* to refer to new information gathered in this field; he has used this term in all his works because by applying general dimensional analysis one can find the values of all quantity exponents, whatever the number of correlated variables.

General dimensional analysis provides the answer to the question: could we derive the values of all quantity exponents from a typically monomial physical relation by using only dimensional analysis calculations? To find an answer I have had to go through several stages: a critical consideration and formulation of several remarks about restricted dimensional analysis; the adoption of new hypotheses and premises; the use of some algebraic concepts never before applied in this field; working out solutions for the new problems; and checking in practice the results yielded by these solutions. In contrast with restricted dimensional analysis, in general dimensional analysis the measurable variables are distributed from the very beginning between the two members of the monomial physical relation to be derived and unknown exponents are ascribed to all variables; as for the values of exponents — which can only be positive — my option was the integers and I have proved that only minimal, or small (close to the unity) values are implied; measurement data were used to estimate the value of the numerical factor, which means that for the general dimensional analysis practice is the criterion of truth. The ratios derived by this procedure could be called quasi-rational.

In other words, I can state that general dimensional analysis introduces a heuristic procedure which proved fruitful in practice, namely : positive solutions involving minimal or small integral numbers for the Diophantine systems of homogeneous linear algebraic equations expressing the condition of the dimensional homogeneity of physical relations can be found ; this procedure leads to relations of the monomial type.

General dimensional analysis and experiment constitute the dimensional method of studying physical phenomena and at the same time the dimensional method of representing, through relations, the data obtained through measurements ; this method is also suggested for experiment.

The Diophantine systems of homogeneous linear algebraic equations expressing the condition of dimensional homogeneity of physical relations can be used in other fields as well ; they express the conditions of mass conservation in chemistry and the condition of conservation of the number of nucleons and of the electric charge in nuclear physics ; and this could be a proof of the unity of these three fields of science.

The problems of general dimensional analysis are dealt with in the works published by the author after 1953, as well as in the present book. The applications of restricted and general dimensional analysis — these are comparatively numerous in this book — are based on information extracted from similar publications, and from the author's own experience. The method of general dimensional analysis leads to an algorithm allowing the physical relations to be derived by means of the electronic computer.

C. I. STAICU

Bucharest

Note

In the List of References at the end of the book a few titles are included which were not mentioned in the text. This is due in the first place to the fact that for the restricted dimensional analysis — i.e. the old dimensional analysis — only the more recently published works were used; secondly, the general dimensional analysis — i.e. the new dimensional analysis — being a novelty, the author displayed a particular care in investigating all the works published around the world on this subject so as to be sure that the new concepts he suggests are really genuine.

The numbering of the tables, figures and footnotes consists of two numbers: the first one represents the number of section (twenty five sections) and the other, the number of order within a section. This system of numbering is used for references too, where the first number represents the chapters of the book (five chapters).

A few mathematical symbols have several meanings; these symbols were built up in the usual way; the initial letter of the meaningful word, in alphabetic order etc.; thus, symbol r may represent either the physical quantity of radius, or a quantity exponent or else the rank of a matrix. For ease of identification, those symbols that are specific to dimensional analysis and have a high recurrence have been marked by an asterisk, which eliminates the need of tiresome explanations, especially when several words are needed for the purposes as, for instance, the number of physical quantities describing a physical phenomenon m^*, the number of equations expressing the condition of dimensional homogeneity of a physical relation e^* etc.

Contents

II. *Restricted Dimensional Analysis*

III. *Applications of the Restricted Dimensional Analysis*

Mathematical Symbols

a Exponent of physical quantity, or of non-dimensional product of powers ; physical variable index

A Physical measurable variable ; physical quantity, characteristic constant, universal constant

$[A]$ Dimensional equation of the physical variable A

d^* Number of fundamental quantities of a consistent system of units

e^* Number of independent linear algebraic equations relating the dimension to the quantity exponents expressing the condition of dimensional homogeneity of a physical relationship

e_k^* Number of equations having the same amount of terms

$°E$ Engler degree for the measurement of fluids viscosity

f Function of physical variables

i^* Number of non-dimensional products of powers

j^* Number of physical quantity exponents which can be additionally determined by using the rule for the resolution of Diophantine systems of equations

k Numerical factor of proportionality ; it is a constant, non-dimensional factor

l_a Dimensional exponent of fundamental quantity L in the dimensional equation of the physical variable A

L Fundamental quantity of length

m^* Number of variables describing a certain physical phenomenon, or taking place in the running of a certain physical phenomenon

m_a Quantity exponent of fundamental quantity M in the dimensional equation written for the physical variable A

m_1^* Number of physical variables in a given function of variables whose definitions or physical relations are known

m_j^* Number of variables the dimensional equations which include a certain fundamental quantity

m_L^* Number of variables the dimensional equations of which include the fundamental quantity L

M	Fundamental quantity of mass
n^*	Number of unknowns in the equations expressing the condition of dimensional homogeneity
p^*	Number of definition relations added to the number of physical relations known between a small number of physical variables of a given function of variables
r^*	Rank of a matrix
t^*	Number of terms in those equations which express the condition of dimensional homogeneity
t_a	Dimensional exponent of fundamental quantity T in the dimensional equation of physical variable A
t_j^*	Number of terms of that equation that expresses the condition of dimensional homogeneity corresponding to a certain fundamental quantity
t_L^*	Number of terms of that equation that expresses the condition of dimensional homogeneity corresponding to fundamental quantity L
T	Fundamental quantity of time
θ	A conversion factor to account for the change in the value and unit of measurement of the thermodynamic temperature
θ_a	Dimensional exponent of fundamental quantity Θ in the dimensional equation of physical variable A
Θ	Fundamental quantity of thermodynamic temperature
λ	A conversion factor to account for the value and the unit of measurement for length
μ	A conversion factor to account for the value and unit of measurement for mass
ν	Kinematic viscosity
ν_j^*	Rate of occurrence of a certain fundamenta quantity in the dimensional equations of physical variables
ν_L^*	Rate of occurrence of fundamental quantity L in the dimensional equations of physical variables
Π	A non-dimensional product of powers
τ	A conversion factor for the value and the unit of measurement for time
φ	A function of non-dimensional products of powers
q^*	Number of fundamental units of measurement in a consistent system of units

Note: In the book some symbols have more than one meaning; the author has indicated when this occurs.

Introduction

Dimensional analysis is the study of relations which describe physical phenomena. It relies on the condition of dimensional homogeneity — which should be satisfied by all rational and empirical relations. This condition refers to the requirement that all terms of a physical relation should be homogeneous, meaning that they should have the same units of measurement, i.e. the same fundamental quantities and the same powers of fundamental quantities. The dimensional homogeneity of physical relations is a strict necessity : indeed, it secures their invariability against changes in the system of units, a change which often occurs in practice. The condition of dimensional homogeneity is expressed through systems of linear algebraic equations between dimension and quantity exponents.

Particularly important in this field is the fundamental theorem of dimensional analysis which refers to the expression, through monomial relations and as a function of the fundamental units of measurement, of the derived units.

According to the restricted dimensional analysis—i.e. the old dimensional analysis — as known at present, the condition of dimensional homogeneity leads to either determinate or indeterminate systems of algebraic equations. In simpler cases, the systems of equations are determinate and hence accept only one solution ; calculations conducted according to the restricted dimensional analysis result in monomial relations between physical variables, which can be solved up to the numerical factor. More often than not, the systems of equations are indeterminate and acccept one general solution ; calculations conducted according to the restricted dimensional analysis lead to relations between non-dimensional products of powers. In such cases, in order to find the values of the numerical factor and of the exponents of non-dimensional products of powers, one has recourse to data gathered from actual practice and, sometimes, to the theory of phenomena.

In restricted dimensional analysis the Π theorem, which refers to the reduction in the number of measurable variables describing a physical phenomenon, is essential; equally important is the fundamental theorem referring to the monomial form of the relation between these variables.

The general dimensional analysis — i.e. the new dimensional analysis — developed by the author, applies to those cases for which the restricted dimensional analysis leads to indeterminate systems of equations between the dimension and the quantity exponents. These cases are solved by the new methodology according to which, in a first stage, the measurable physical quantities are distributed between the two members of the monomial relation to be derived and unknown exponents are assigned to all quantities; then, the homogeneous, linear algebraic equations expressing the condition of the dimensional homogeneity of the physical relation are written and the equations of the truncated triangular system, which generally results, are solved in the increasing order of the number of terms; since the system of equations is a Diophantine one, positive, integral, minimum values are ascribed to the unknowns, leading to values of the like nature for the remaining unknowns; on the basis of experimental data, the non-dimensional numerical factor is computed; this factor should be practically unchanged in value over the whole range of definition.

The results obtained by general dimensional analysis computations are conditioned by the nature of functions of the measurable variables considered when describing the physical phenomena. When the functions are inaccurate, positive, integral, small (close to unity) values should be chosen for the quantity exponents and several trials may be necessary before it is possible to establish a relation.

In general dimensional analysis, the theorem of the general dimensional analysis referring to the monomial form of relations is essential; this theorem differs from the homologous one used in restricted dimensional analysis. Equally important is the theorem of positive, integral, minimum or small values ascribed to quantity exponents in order to find their values by using dimensional analysis computations alone.

The positive, integral, minimum values ascribed in the general dimensional analysis to the quantity exponents of typically monomial physical relations appear in chemistry, as values of the substance coefficients in chemical equations, as well as in nuclear physics, as values of coefficients of the nucleus and of the elementary particles included in the equations written for nuclear reactions.

Dimensional analysis — either restricted or general — applies to all fields of physics and technical sciences where measurable variables are used. The examples presented in this book are illustrative of the large majority of problems dealt with. The ultimate goal of the applications is to obtain correct relationships between physical quantities or non-dimensional products of powers; some of the applications are intended to serve as a check for the relations derived by other methods.

I. The Basis of Dimensional Analysis

1 Entity, Physical Quantity, Measure

1.1 Definitions

Physical phenomena can be described by a finite, comparatively small number of distinct concepts: length, velocity, amount of heat, etc. called entities. These are implied in any scientific statement or physical experiment. Each entity has some independent properties; in dimensional analysis however, only the *PHYSICAL QUANTITY* of the *ENTITY* is important.

Physics is concerned with entities that are characterized by observable effects. The description of such an effect represents the qualitative definition of the entity, which is prior to and independent of the laws governing physical phenomena. Such definitions are useful in order to know the particular entity implied in a discussion and also to recognize any observable entity, when met.

The expression currently used for entity is physical quantity, the latter being, in fact, a property of the former, as stated above. However, this substitution is dangerous when drawing conclusions about the nature of entity, for, if an object is replaced by its properties, difficulties can arise. Consequently, this substitution can be accepted only for the sake of simplification and it should never be forgotten that what is really meant is the entity's physical quantity, i.e. the physical quantity of length, the physical quantity of force, etc. [1.8].

Regarding this expression, Leibniz recommends the use of any numerical figures or other figures known by using integers which are required in order to distinctly recognize a physical quantity; thus, from the continuous physical quantity we come to discrete measures[1.1]. Even Kant said that nobody can define the concept of physical quantity in general, except by using an example of this kind : the physical quantity is the determination of a factor which allows one to realize how many times the unity is contained in this factor[1.2] [1.21].

It can be said that two observable entities, A and B, are comparable when there is an universal, operational definition of the ratio

$$\frac{A}{B} = n \tag{1.1}$$

where n is a pure number, showing that entity A is proportional to entity B. "Operational" refers to the fact that in order to compare entities A and B both the instruments used and the operations carried out should be known. The condition of "universality" requires that the entity ratio be independent of the nature of the materials used in building the above instruments.

However, it is not always possible to compare entities. Take for instance the feeling of joy : one can, of course, say that some feelings are more intense than others; but no instrument has yet, been invented, to measure how many times one joy is greater than another. The same applies to beauty, kindness, admiration, etc.

The definition of the ratio of two observable entities immediately leads to the definition of equality and summation. Indeed relation

$$\frac{A_1}{A_0} = n_1 \tag{1.2}$$

leads to equality :

$$A_1 = A_0 \text{ when } n_1 = 1; \tag{1.3}$$

1.1 Leibniz, G. W., *Nouveaux essais sur l'entendement humain,* Book II, Chap. XVI, p. 112.

1.2 Kant, I., *Transcendental Analytics,* Book II, Chap. III.

and relations

$$\frac{A_1}{A_0} = n_1, \qquad \frac{A_2}{A_0} = n_2, \qquad \frac{A_3}{A_0} = n_3 \tag{1.4}$$

lead to the sum

$$A_1 + A_2 = A_3 \text{ when } n_1 + n_2 = n_3. \tag{1.5}$$

It is considered that in a sequence $A_0, A_1, \ldots$ of observable entities, these entities can be compared at least two-by-two. But there are different *MEASURES* of a given *PHYSICAL QUANTITY*. For example: the distance between two cities, the width of a river, the height of a building are all *MEASURES* of the physical quantity of length; the durations of an athletic competition, of a school lesson and of pendulum oscillations are *MEASURES* of the physical quantity of time.

Physical quantities may be considered as abstract concepts when the measure of their value is taken into account. This agrees with the natural passage from concrete concepts (measures) to abstract concepts (physical quantities). According to some authors [1.32], physical quantities have an objective existence; such theories however, lack interest.

The distinction between physical quantity and measure is particularly useful when perfectly clear ideas are sought. In current language however, even in physics, the word "physical quantity" is currently used and only occasionally the word "measure". The explanation lies, of course, in the fact that a general concept is used for the particular. This also happens in physics [1.32].

1.2 Primary and Secondary Physical Quantities

Some entities are immediate. They have an intuitive character. Their properties are easy to establish. The most important is the property which refers to the physical quantity of an entity: for instance, length, time, etc. Other entities can be taken into consideration only if problems of geometry, of physics, or of other scientific disciplines are solved first; such entities are, for instance, velocity, viscosity etc. Entities of the first category are called primary physical quantities (or entities), while those of the second category are called secondary physical quantities. This classification of physical quantities is of interest in dimensional analysis.

Primary physical quantities are introduced according to needs; the relevant measuring unit should also be specifically chosen, and the measuring procedure should be implicitly established. In any science only the minimum of primary physical quantities needed are used; this minimum is so chosen that no primary physical quantity can be defined as a function of other primary physical quantities; meanwhile, all physical quantities used in one specific field of science should be so related to each other that any one could be defined as a function of any other. In physics, the number of primary physical quantities should be high enough to afford the description — using both primary and secondary physical quantities — of all physical phenomena; on the other hand, the number of primary physical quantities should be small enough to permit analytical relationships among them, for certain given measuring units.

Secondary physical quantities are defined by means of primary physical quantities; their values are given by a function of values of these primary physical quantities which apply to the specific field to which the secondary physical quantity considered belongs. Take velocity, for instance: it is given by the quotient of dividing the distance travelled by the time of travel; and it should be stressed that this is not a fact which could be experimentally checked. "Velocity" is merely called the result of the mathematical operation mentioned above.

The only difference between primary and secondary physical quantities is the fact that the former are experimentally established; that is they are synthetic in character, while the latter are defined through equations, and are analytical in character. As can be seen, primary physical quantities are not arbitrary.

The relation equating the value of a secondary physical quantity to the value of the primary quantity function defining a secondary physical quantity is called the secondary physical quantity relation of definition. In secondary physical quantity relations of definition, coefficients may intervene to which simple or important values are assigned. For example, in the relation defining kinetic energy E:

$$E = \frac{1}{2} mv^2 \tag{1.6}$$

where m is the mass and v is the velocity, the coefficient 1/2 appears.

The quality of the primary or secondary physical quantity is a relative one. Physical quantities used to define a secondary physical quantity can be, themselves, secondary physical quantities, previously defined. In principle, there is no restriction on the possibilities of definition of a secondary physical quantity; indeed, any function of primary physical quantities, as well as any secondary physical quantity previously defined, can be used. However, to be useful, any new definition of a secondary physical quantity should bear a physical meaning, that is it should lend itself to being represented in the mind.

Primary and secondary physical quantities are therefore classified according to how they are introduced in physics.

It should be mentioned that, till now, no specification has been made as to the form of the secondary physical quantity relation of definition. This will be dealt with later, when a more comprehensive discussion of the subject will be presented (Section I.5, p. 44).

1.3 Variable and Constant Physical Quantities

Physical quantities intervening in a given physical phenomenon may be either variable or constant. At this stage of our knowledge, a high enough number of variable physical quantities are available; constant physical quantities are, however, very few. As a consequence, for any physical quantity connected to any well-determined physical phenomenon the simple designation of "variable" is used. Physical quantities may be dependent or independent. Constant physical quantities may be either characteristic or universal.

This will be best explained by means of an example. Suppose we have a wire stretched by external forces; in such a case, the intervening physical quantities are the wire length l, the wire cross-section S, the force F; and the wire elongation Δl as a result of the external force. Now, applying Hooke's law, we can write the following relation

$$\Delta l \sim \frac{lF}{S} \tag{1.7}$$

where the sign $\sim$ stands for the proportionality relationship ; between the values of the above physical quantities, we can write the following relation

$$\Delta l = k\frac{lF}{SE} \tag{1.8}$$

where E is Young's modulus. Physical quantities l, S and F are the independent variables while Δl is the dependent variable. Young's modulus E is a characteristic constant physical quantity (for a particular material) ; its measure depends on the wire properties and on the units of measurement assigned to length l, cross-section S and force F.

In other cases, the values of constant physical quantities are independent of the nature of the body ; these are called universal constants. Thus, for instance, in the relation between gravitational mass m_g and inertial mass m_i, which is of the form

$$m_g = \sqrt{G}\cdot m_i, \tag{1.9}$$

the gravitational constant G intervenes. Other examples of universal constants are the velocity of light, Wien's constant, etc. Their values depend on the units of measurement used.

Universal constants represent mathematical (analytical) expressions for the fundamental properties of matter. Regarding their significance in science and the important part they play in any scientific knowledge, Planck said that universal constants were not invented for practical convenience ; on the contrary, they evolved by themselves, owing to the agreement they brought about in the results of all important measurements and (this is the essential point) because we know beforehand that all future measurements will accurately lead to the same constants [1.3]. Speaking about the same constants, Heisenberg says that universal constants are determinant for the natural scale ; they are characteristic physical quantities which cannot be reduced to other quantities [1.4] [1.12].

[1.3] Plank, M., *Autobiographie Scientifique*, Al. Michel, Paris, 1960, p. 208.
[1.4] Heisenberg, W., *Physique et Philosophie*, Al. Michel, Paris, 1961, p. 171.

2 Units of Measurement

2.1 Measurement of Physical Quantities

The physical quantity is the expression of an important property of the entity; it may be either variable or constant. Physical quantities which differ only in measure are of like nature. The measure of a physical quantity is expressed by a number, which is obtained as a result of a measurement. A measurement will establish how many times a physical quantity will be contained within a given quantity of the like physical nature. The quantity used in measurements is called a unit of measurement. A body representing a unit of measurement is a constant physical standard called the standard of the physical quantity.

The unit of measurement U_A is a definite measure A_0 of a quantity, i.e. $U_A = A_0$. It can be adopted for each physical quantity. Hence, writing for several physical quantities A_1, A_2, . . . the ratio

$$\frac{A_1}{U_A} = a_1, \qquad \frac{A_2}{U_A} = a_2, \ldots \tag{2.1}$$

for each measure of the physical quantity A considered, a number a will result, which is called the value of the physical quantity having U_A as a unit of measurement; this value is established by measurement.

Let us assume that the unit U_A is replaced by another unit, U'_A; then, another value for the physical quantity a will be obtained, namely

$$\frac{A}{U'_A} = a'. \tag{2.2}$$

Now, relation

$$\frac{a}{a'} = \frac{U'_A}{U_A}, \tag{2.3}$$

according to which *the quotient of the values of the same physical quantity is equal to the reciprocal of the units ratio*, is evident. According to [1.4] [2.1], relation (2.3) is the expression of the fundamental theorem of the units of measurement.

Measurement of primary entities are a simple matter : a given, say, length is compared against a unit of length. For secondary entities, a direct comparison against the adequate unit is not always practical or convenient; in such cases, one has to look for another way to turn the difficulty. Therefore, use will be made of the secondary physical quantity relation of definition, leading to relation

$$a = f(a_1, a_2, \ldots) \tag{2.4}$$

in terms of which the value a assigned to the physical quantity of a new entity, is expressed as a function of values $a_1, a_2, \ldots$ of physical quantities of other known entities; these can be of like nature (say, lengths) or of different natures (say, lengths, time, etc.).

Referring to relation (2.4), it should be specified that the equality sign is valid only between measures of like nature. Also, function f expressing certain mathematical operations is significant only when in the relevant operations numbers (not entities) are involved. For exemplification purposes, let us assume that the concepts of force and acceleration were established and that one knows how to measure the corresponding physical quantities. Experiment has shown that to a given body, driven by a variable force, a variable acceleration is imparted. If we compute the quotient of numbers which are measures of the physical quantities of force F and of acceleration a, we shall see that this quotient always has the same value; hence we may write :

$$\frac{\text{the number giving the measure of force}}{\text{the number giving the measure of acceleration}} = \text{const.}$$

A new entity of a constant physical quantity is thus derived; it was agreed to call it "mass". The mass concept is therefore introduced in physics by means of the definition relation

$$\frac{F}{a} = km \tag{2.5}$$

where k is a numerical proportionality factor. Relation (2.5) shows the operation to be carried out; this operation should include only measures of force, acceleration and mass.

[2.1] p. 34.

The measuring operation leads to a bi-unique correspondence between the physical quantities of an entity and a sequence of real numbers; thus the operating elements of arithmetic are obtained — numbers — and consequently arithmetic laws can be applied. This correspondence is, however, arbitrary since it depends on the agreement on which the choice of the units of measurement relies.

Planck correlates physical quantity measurements with universal constants and emphasizes the fact that the existence of the universal constants is the tangible proof that nature holds something real, which is independent of any measurement process [2.2] [1.14].

Appendix 1 is a summary of the problems of the unit of measurement as applied in geometry to achieve beauty.

2.2 Fundamental and Derived Units of Measurement

As a rule, some entities are considered fundamental. Such entities are established with due consideration given to the convenience of the comparison operations to be carried out. Arbitrary units of measurement are assigned to the entities thus chosen, since in such cases nothing else is required; these become fundamental units of measurement. It follows therefore, by definition, that no relationships exist among the fundamental units of measurement.

In Euclidean geometry, length, for instance, is a fundamental entity and an arbitrary unit is chosen for length; it could be the centimetre, but other units could do as well, e.g. the metre, the foot, etc. In mechanics, the most frequently adopted convention considers as fundamental the following entities: length, mass and time. For the fundamental entity called mass, the variety of fundamental units is even greater.

In principle, an arbitrary unit could be assigned for the measurement of each quantity; in such a case, the units thus established would all be fundamental. However, there is no advantage in handling only fundamental units. To cut down the number of fundamental units of measurement derived units were introduced. By analogy entities corresponding to the derived units are called derived entities. Derived units are obtained from the relations of definition used for the secondary physical quantities.

[2.2] Planck, M., *Autobiographie Scientifique*, Al. Michel, Paris, 1960, p. 208.

No conditions are imposed on the choice of the units of measurement. According to [1.8] [2.3], fundamental units would be more aptly called "reference units" by analogy with the reference axes used in geometry. Each group of fundamental units has certain qualities, as has each system of reference axes.

The classification of units of measurement brought about, implicitly, a new classification of entities or, bearing in mind the looseness of language (Section I.1), one can say that a new classification of physical quantities appeared, namely fundamental and derived. Here, physical quantities are classified from the standpoint of the units of measurement.

It should be stressed that the two classifications established for the physical quantities should not lead to confusion. Primary physical quantities should not be taken as fundamental. As already shown, the primary physical quantities are direct, intuitive concepts and no relationship exists among them ; on the other hand, the fundamental physical quantities are arbitrarily chosen and form the basis of a system of units of measurement. The fact of being or not being a primary physical quantity is a quality inherent in each quantity, while any primary or secondary physical quantity may be a fundamental physical quantity. Secondary physical quantities, in turn cannot be taken for derived physical quantities. Indeed, the fact of being or not being a secondary physical quantity is, likewise a quality inherent in each quantity, whilst the fact of being a derived quantity is the result of an agreement.

It was suggested that all fundamental physical quantities should be primary physical quantities ; however, this is not always desirable [1.44].

2.3 Choice of Fundamental Units of Measurement

In geometry, by experiment and judgement we can show that the value S of the area of a rectangle is proportional to the product of numbers resulting from the measurement of rectangle sides l_1 and l_2. Hence we can write

$$S = kl_1 l_2 \tag{2.6}$$

[2.3] p. 7.

where k is a proportionality factor. The value of factor k depends on the units of measurement adopted ; this factor is needed to establish a relation between the units of the two physical quantities.

Factor k might take any random value, for instance, 1, 2, 5, e, $\sqrt{\pi}$ etc. For the sake of simplicity, it is best when this factor is made equal to unity ; as a consequence, relation (2.6) becomes :

$$S = l_1 l_2. \tag{2.7}$$

If relation (2.7) is adopted, we cannot choose both units of measurement at random, because one unit is derived from the other, arbitrarily selected one. If, for instance, we choose the unit of length as fundamental, the unit for area becomes a derived unit and will be equalled by the area of a square whose side is the unit of length. This is what is now happening in geometry : the fundamental unit is a measure of length and all other geometrical units are derived units. Should the unit for area be adopted as fundamental in geometry, the correlation between the two units of measurement would be established in a similar manner.

In theory, there is no argument against adopting other values for factor k. As already stated, the choice $k = 1$ is an arbitrary one. This particular case of the agreement cannot lead to a formula expressing a relation of necessity between the relevant physical quantities. The unit value adopted for the proportionality factor was imposed by practical considerations and in particular by those referring to the size of the derived units.

In conclusion, we can state that : *when in a relation describing a physical phenomenon entities are implied whose units of measurement are already established (either fundamental or derived), this relation will compulsorily include a proportionality factor — a numerical conversion coefficient, and any other relation derived from this relation will also include such a coefficient* ; this statement is generally valid [1.8].

Whilst the choice of units is an arbitrary one, the above rule should nevertheless be borne in mind when such units are established. As a consequence, the selection of units is dictated by opportunity considerations. The most suitable fundamental entities are those among which the relations are least often used in practice ; indeed, in such a case, the conversion coefficient will be used only as an exception.

When choosing the fundamental entities, due consideration should be given to their capability of permitting the build-up of physical standards which are invariable in shape, thus affording measurements of the highest attainable accuracy. From this point of view, the units for length, mass, time and force are most advantageous.

2.4 The Consistent System of Units

Physical phenomena are described by means of a small, finite number m^* of physical quantities of different natures. It is considered that for $m_1^* < m^*$ physical quantities of like nature, either their definition equations or some relations representing physical laws are known; the number of these relations is denoted by p^*. Consequently, we might choose at random

$$q^* = m^* - p^* \tag{2.8}$$

physical quantities whose units are also arbitrary. The q^* arbitrary units will be fundamental and will serve to derive, from the p^* relations, the derived units.

The whole set, consisting of the fundamental units and of those derived from them, forms a consistent system of units of measurement. Several consistent systems of units are known; they differ either in the nature of the fundamental physical quantities, or in the measure arbitrarily chosen as a unit. Fundamental physical quantities of length, mass and time have served as a basis for two systems of units, i.e.: CGS and MKS. In technics, the MKfS system is often used.

Referring to the problem of units of measurement, Heisenberg, after presenting the two universal constants of relativity and of quantum theory respectively — i.e. light velocity in vacuo and the quantum of action — suggests that there must be a third universal constant in nature, for at least three fundamental units are needed to build up a complete system of units. This can be easily ascertained by considering, say, the use by physicists of the CGS system which is, after all, a mere agreement: one unit of length, one of time and one of mass are enough to build up a complete system, but there should be at least three such units. Likewise, we could substitute for these units another set of three, say length, velocity and mass ; or else length, velocity and energy. The velocity of light and Planck's constant are but two of such units. We therefore must have a third one, and only a theory containing this third unit could be operational in determining the mass and other properties of the element particles[2.4]. Heisenberg considers that the third universal constant could be a universal length inferior to the nucleic radii of light atoms, perhaps of the order of 10^{-13} cm [1 · 14][2.5].

Some countries have currently adopted the International System of Units (SI). It relies on seven fundamental physical quantities as shown in Table 2.1.

TABLE 2.1. **Fundamental physical quantities and units in the International System (SI)**

Fundamental physical quantity	Measuring unit	
	Designation	Symbol
1. Length	metre	m
2. Mass	kilogramme	kg
3. Time	second	s
4. Intensity of electric current	ampere	A
5. Thermodynamic temperature	kelvin	K
6. Intensity of light	candle	cd
7. Quantity of substance	mol	mol

As can be seen from relation (2.8), the number of fundamental units is a definite one ; their nature, however, is not definite. As already shown, the choice of fundamental units is governed by both practical and theoretical consideration.

Euclidean geometry is, in the first place, concerned with lengths and one single fundamental physical quantity is used ($q^* = 1$) ; its unit is arbitrarily chosen. In kinematics, one also has to account for time, while in statics we have to bear in mind force ; hence both fields use two fundamental physical quantities ($q^* = 2$). In dynamics, the number is three ($q^* = 3$) and can be either length, mass and time, or length, force and time. To these fundamental physical quantities a variety of units correspond such as : the metre, the kilogramme-mass, the second, the kilogramme-force.

2.4 Heisenberg, W., *Physique et Philosophie,* Al. Michel, Paris, 1961, p. 190.
2.5 p. 109.

In other fields of science, new fundamental physical quantities are introduced as, for instance: intensity of the electric current in electricity; temperature in thermodynamics; and light intensity in the study of light phenomena. The corresponding units are: the ampere, the kelvin and the candle respectively. Consequently, in a problem of physics where these six fundamental units are implied, $q^* = 6$.

Number q^* has neither an absolute nor a necessary value, in any scientific field. Thus, for instance, in dynamics usually $q^* = 3$, since the definition of the length, mass, time and force is convenient and any measurement carried out on that basis is accurate enough. However, in dynamics the number of fundamental units could be dropped, or even reduced to zero.

Let us consider the system of three fundamental units currently applied in mechanics and relying on the physical quantities of length, mass and time; within this system of units, the force is defined by the fundamental law of mechanics

$$F = m \cdot a, \tag{2.9}$$

while the law of universal gravity is written in the form:

$$F = k' \frac{mm'}{l^2}. \tag{2.10}$$

This relation includes the factor k', which is a universal constant whose measure univocally depends on the arbitrary physical quantity of fundamental units. If the law of universal attraction is to be written in the form

$$F = \frac{mm'}{l^2}, \tag{2.11}$$

the fundamental units cannot be arbitrary any more; their values should be chosen so that when computing the universal constant one should reach the value $k' = 1$. Under these conditions, the right-side members of relations (2.9) and (2.11) can be equated and thus:

$$m' = al^2. \tag{2.12}$$

This relation shows that mass is not to be considered a fundamental physical quantity, as long as it can be defined by means of the physical quantities of length and acceleration. Hence, its unit can be derived from the length and time units, which remain arbitrary. It follows that the number of fundamental units has dropped by one. This new system is called the astronomical system of units.

One can go even further with the reduction of the number of fundamental units. To this end, a constant should be considered. The velocity of light in vacuo is constant. Now, it can be so arranged that in the astronomical system of units this constant takes the unity value, i.e. light velocity can be accepted as a unit for velocity. In such a case, the second, for example, could be kept as a fundamental unit of time and $3 \cdot 10^5$ km could be taken as a unit for length. As a consequence, the length unit becomes a derived unit.

Even though the number of fundamental physical quantities is reduced to unity, there still remain universal constants to allow for a further reduction. Consider, for instance, atomic structure: Planck has equated to unity the constant which bears his name. As a consequence, the unit for length cannot be chosen at random and the result will be $q^* = 0$, i.e. a system of units with no fundamental unit; all fundamental units have vanished.

However, such systems of units, as well as those suggested by Hartree, Rouarc and other physicists who equated to unity the various universal constants, are not used in technics.

It is worth mentioning that since the choice of both the fundamental physical quantities and the fundamental units is arbitrary, one should have the guarantee that the changes implied will not lead to errors; this situation is handled by a correlation to be dealt with elsewhere (Section I.6).

3 Relationships and Laws in Physics

3.1 Definitions and Classifications

Any physical phenomenon is described by a finite, small number of parameters. Consider, for instance, the motion of a pendulum within a uniform gravitational field. Pendulum kinetic energy, or impulse, depends on four independent variables, namely pendulum length l, the normal acceleration in free fall under gravity g, the angular deviation from the vertical α and the pendulum mass m. The pendulum linear velocity, or acceleration, depends on only three independent variables, namely l, g and α. The pendulum period is a function of only two independent variables : l and g. It follows that the pendulum motion can be studied by means of four independent variables l, g, α and m ; it further follows that some physical quantities are functions of a lower number of variables.

Relationships between physical quantities can be of various kinds. The most general are called laws. Others are called theorems ; these are established by mathematical deductive reasoning. The number of laws has been reduced since the purpose of theoretical physics is to establish laws so general that some of them can become theorems. The definition relationships of physical quantities are another group of links among physical quantities. Some relationships among physical quantities form a group adapted to certain measuring units.

In any physical law the measures imply that in the course of a phenomenon changes are occurring. The law expresses a state of the phenomenon, or the way the measure of a physical quantity, characteristic of the evolution of the physical system in certain given conditions, will change as a function of the variable measures of certain physical quantities. These laws have no specific designation. Other laws — called conservation laws — express a restriction of the conversion possibilities. According to these laws, only those conversions are possible in which the value of a certain physical quantity is preserved. The laws of conservation can be general — i.e. applicable in all conditions — or of a restricted applicability. The most general are : the law of energy conservation and the law of electric charge conservation. Conservation laws with restricted applicability are met in nuclear physics and imply properties which have no equivalent in macroscopic physics [1.39].

There are, in many fields of physics, phenomena whose laws are expressed through relationships of the same form ; the specific character of each phenomenon appears from the meaning of variables and numerical coefficients in the given relationships. Thus, for example, Fourier's law of thermal conduction, Fick's law of diffusion, and Newton's law of internal friction all have the same form. The identity of relationships expressing phenomena pertaining to the different fields of physics allows a phenomenon taking place in a field where experiment is more difficult, to be studied by observing another phenomenon taking place in another field, more accessible to experimental investigation. These analogies lead to simulation, a technique frequently used in physics and technology.

The laws of physics can be either general or specific concerning a particular material. Thus, for instance, the fundamental law of dynamics (2.9) is a general law. The expression of this law, $F = m \cdot a$, conveys no indication of the particular material of the body of mass m which, under the action of force F, is moved with an acceleration a. In the expression of these laws, universal constants appear [1.32]. Ohm's law establishing the relationship between the intensity of the electric current I, the potential difference U and the resistance R, is expressed

$$I = \frac{U}{R} \tag{3.1}$$

and is a material law since the resistance R depends on the nature of the conductor material. Material laws include characteristic constants [1.32].

Let us consider that for the description of a physical phenomenon we have used physical quantities $A, B, C, \ldots$ and that this particular phenomenon is governed by a law. In the most general form, we may write the relationship

$$f(A, B, C, \ldots, M, N, P, \ldots) = 0 \tag{3.2}$$

or, putting this relationship into an explicit form:

$$P = f(A, B, C, \ldots, M, N, \ldots). \tag{3.3}$$

Any relationship of this type (3.2), (3.3), connecting physical quantities is a mathematical relation. Such a relationship can be converted into a physical one, that is into a relation between measures of the physical quantities. As is known, this is done through a measurement process.

3.2 Relation Conversion

Conversion of mathematical into physical relations will be described by means of an example. Let us consider Newton's mathematical relation expressing the known meanings, i.e. force, mass and acceleration respectively. To be practical, physical quantities F, m and a implied by this relation should be measured. The units of force, mass and acceleration are noted by U_F, U_m and U_a respectively. In terms of (2.1), the measures of these physical quantities can be written by adding to their symbol the subscript v, thus:

$$F_v = \frac{F}{U_F}, \quad m_v = \frac{m}{U_m}, \quad a_v = \frac{a}{U_a}. \tag{3.4}$$

From (2.9) and (3.4), it follows that

$$F_v U_F = m_v \cdot U_m \cdot a_v \cdot U_a \tag{3.5}$$

or

$$F_v = \frac{U_m \cdot U_a}{U_F} \cdot m_v \cdot a_v, \tag{3.6}$$

from which the known relation

$$F_v = k \cdot m_v \cdot a_v \tag{2.5}$$

is derived. This, it should be emphasized, refers to the measures of physical quantities. Information on factor k was obtained on this basis; it is defined from relation:

$$k = \frac{U_m \cdot U_a}{U_F}. \tag{3.7}$$

The proportionality factor represents a parasite numerical coefficient whose value depends on the measuring units of the physical quantities shown in (2.10). The conclusion has a general validity and can be stated thus: *any conversion from mathematical into the corresponding physical relationship brings about a parasite coefficient the value of which depends on the units of the physical quantities implied in the given relation.*

The presence of a parasite coefficient in the physical relations is the cause of their somewhat complex structure. Therefore, it is necessary to eliminate, as far as possible, such coefficients from these relations. To this end, the units of physical quantities are chosen so as to equal coefficient k with unity. Thus, relation (3.7), expressing the parasite coefficient, becomes a conditioning relationship. By eliminating the parasite coefficient from the physical relations, the number of fundamental units which can be arbitrarily defined is greatly reduced. They serve as an underlying basis for defining derived physical quantities. This has already been dealt with (Section I.2).

An important problem, which was not been dealt with as yet refers to the form in functions (3.2) or (3.3) of the physical quantities describing a studied phenomenon. This will be dealt with elsewhere (Section II. 8,9).

3.3 General Form of a Physical Relationship

Let us consider the implicit, most general form of a physical relationship (3.2). There is need to particularize the $m^* - 1$ physical quantities which can be independent variables. Any of the quantities can be made explicit and, as a consequence, the set of $m^* - 1$ physical quantities left on one side of the expression plays the part of the independent variables. This was how (3.3), for example, was written.

The explicit physical quantity P, which might be either primary or secondary, is expressed through a function of fundamental physical quantities, for example by relations of the form

$$P = f(\mathrm{L}, \mathrm{M}, \mathrm{T}) \tag{3.8}$$

where the fundamental physical quantities admitted were length L, mass M and time T. Likewise, physical quantities of the second member which might be either primary or secondary are expressed through relations of the form (3.8), i.e.

$$f(A, B, C, \ldots) = \varphi(\mathrm{L}, \mathrm{M}, \mathrm{T}). \tag{3.9}$$

It follows therefore that the most general form of any physical relationship expressing a physical law is

$$f(L, M, T) = \varphi(L, M, T). \tag{3.10}$$

3.4 The Numerical Factor

Any physical law is the expression of a relationship established between several physical quantities. Such a relation always implies those physical quantities which describe a definite physical phenomenon. Other relations include, in addition to physical quantities, a numerical factor. Its value is independent of the units used.

In any physical relation, the numerical factor stands for the proportion of the measure of the dependent variable to the value of the monomial built up with the remaining variables in the relevant relation. The characteristic feature of any numerical factor is that, generally, its value remains constant over the whole range of measures which the physical quantities of the relevant physical relation can take. There are, however, exceptional cases when the value of the numerical factor, while staying constant over restricted portions of the range, takes two or three different values over the whole range.

The numerical factor in the relations derived through the analytical method is often due to the fact that such relations are established by one or several integrations; in such cases, the value of the numerical factor is established by means of initial and limited conditions.

When theoretical information available is not adequate, the value of the numerical factor can be established by experiment.

According to [1.31][3.1], the numerical factor is a universal constant.

As for the constant value of the numerical factor, as known, physical relations may imply other constants as well, namely: the parasite coefficient due to the units, the coefficient implied by the definition relations worked out for some secondary physical quantities, the characteristic constants, and the universal constants. Of course, this wide variety of constants should not be a cause for errors, particularly when they are to be deduced from the unique value of the numerical factor.

[3.1] p. 42.

4 Homogeneity

4.1 Mathematical Homogeneity

In mathematical — and physical — sciences, homogeneity is that property of a physico-chemical system of having the same physico-chemical characteristics in all its points. However, mathematical homogeneity differs from the physical one. Mathematical homogeneity deals with abstract quantities, deprived of any physical meaning. Thus, for instance, when we say that relation

$$ax^2 + xz + b\frac{x^2y}{z} = 0, \tag{4.1}$$

where a and b are numerical constants, is a homogeneous equation, one need not know the significance of each of the variable parameters x, y, z; all that is needed is that each monomial be of the same order. Physical homogeneity, however, is more restrictive and requires that each term be, in addition, of the like physical nature.

Homogeneous functions have a few fundamental properties. In terms of the theory of functions of a real variable, it is said that a function y of n variables

$$y = f(x, y, z, \ldots) \tag{4.2}$$

is homogeneous when, by multiplying all variables by a factor k, the whole function is multiplied by the same factor k raised at a power m; this power represents the homogeneity order. Function (4.2) is homogeneous with respect to physical quantities x, y, z ... if, by their multiplication by a factor k, it follows that

$$f(kx, ky, kz, \ldots) = k^m f(x, y, z, \ldots). \tag{4.3}$$

Relation (4.1), for example, is a homogeneous function of the second order since by multiplying the variables by a factor k the function will be multiplied by k^2, i.e.

$$a(kx)^2 + kx \cdot kz + b\frac{(kx)^2 \cdot ky}{kz} = k^2\left(ax^2 + xz + b\frac{x^2y}{z}\right). \tag{4.4}$$

The homogeneous function is independent from the units. Consider, for instance, the relation expressing the area of a rectangle (2.7); dividing by S results in

$$\frac{l_1 l_2}{S} = 1. \tag{4.5}$$

Relation (4.5) is always valid, irrespective of the measuring unit adopted for the physical quantities l_1, l_2 and S, i.e. the metre or the centimetre.

A homogeneous function of order m may be expressed as the product of any variable raised to a power m by a function of the other variables, divided by the first variable. Let us consider relation (4.3) and let $k = \dfrac{1}{x}$; we may write

$$f\left(1, \frac{y}{x}, \frac{z}{x}, \ldots\right) = \left(\frac{1}{x}\right)^m f(x, y, z, \ldots) \tag{4.6}$$

from which

$$f(x, y, z, \ldots) = x^m f\left(1, \frac{y}{x}, \frac{z}{x}, \ldots\right). \tag{4.7}$$

Including the numerical constant from the second member in the function symbol results in

$$f\left(1, \frac{y}{x}, \frac{z}{x}, \ldots\right) = f_1\left(\frac{y}{x}, \frac{z}{x}, \ldots\right). \tag{4.8}$$

From (4.6) and (4.7) it follows that

$$f(x, y, z) = x^m f_1\left(\frac{y}{x}, \frac{z}{x},\right). \tag{4.9}$$

Relation (4.1), for example, leads to

$$ax^2 + xz + b\frac{x^2 y}{z} = x^2 f_1\left(\frac{y}{x}, \frac{z}{x},\right) \tag{4.10}$$

which agrees with what was said above.

When a homogeneous function includes only variables of like nature, it can be so converted as to include one single variable. Assuming, for instance, that the variables in (4.10) represent three lengths, we can, of course, write :

$$y = \lambda_1 x, \qquad z = \lambda_2 x. \tag{4.11}$$

From (4.1) and (4.11) it follows that

$$ax^2 + x\lambda_2 x + bx^2 \frac{\lambda_1}{\lambda_2} = x^2\left(a + \lambda_2 + b\frac{\lambda_1}{\lambda_2}\right) = k\lambda x^2 \tag{4.12}$$

where λ is a number. Likewise, when a homogeneous function includes simultaneously several lengths, several masses and several times, it may be converted so as to include only one length, one mass and one time.

When a homogeneous function includes one length, one mass, one time..., the relation between them takes the form of a product of powers. This is known as the fundamental theorem of dimensional analysis. As a consequence, in view of its utmost importance, the demonstration of this theorem will be presented in detail in Section I.5.

Any homogeneous function including physical quantities can be reduced to a function with a smaller number of non-dimensional products of powers Π. This is known as the Π theorem and is highly important in dimensional analysis. This theorem will be considered in detail elsewhere (Section II.8).

When a homogeneous function represents a physical phenomenon at a natural scale, it implicitly represents the small-scale model of the same phenomenon. Assuming that a physical phenomenon involves the length L, the time T, the velocity V, the acceleration A and the volumetric rate of flow Q, between these physical quantities we can write the relation :

$$f(L,\ T,\ V,\ A,\ Q) = 0 \tag{4.13}$$

and for the small-scale physical phenomenon

$$l = k_1 L,\ t = k_2 T,\ v = k_3 V,\ a = k_4 A,\ q = k_5 Q. \tag{4.14}$$

From (4.13) and (4.14) it follows that

$$f'(l,\ t,\ v,\ a,\ q) = 0 \tag{4.15}$$

which represents the small-scale physical phenomenon. From what has been said above, a homogeneous function is independent of the unit; hence (4.13) and (4.15) represent the same physical phenomenon.

As a corollary, a natural-scale physical phenomenon will be expressed by the relation

$$\varphi(\Pi_a,\ \Pi_b,\ \ldots) = 0 \tag{4.16}$$

while at a small-scale it will be expressed by the relation

$$\varphi(\Pi'_a,\ \Pi'_b,\ \ldots) = 0, \tag{4.17}$$

and the condition of similitude between (4.16) and (4.17) is

$$\Pi_a = \Pi'_a,\ \Pi_b = \Pi'_b,\ \ldots \tag{4.18}$$

4.2 Physical Homogeneity

It was Fourier who was the first, at the beginning of the last century, to make the important comment that any physical phenomenon is an original datum. It is implied in a definite physical system, independent of man, although this physical system could be built by man.

The first stage of any cognition process of physical phenomena is the description of the phenomenon. To this end, the physical quantities A, B, C, ..., known at a certain stage of the development of scientific knowledge, are used. The relationships among the physical quantities used to describe any physical phenomenon express the physical law of the phenomenon under study.

When it comes to physical quantities, their quantitative apprehension — as a more advanced stage of scientific knowledge — compulsorily requires the use of the measurement process. For the measurement of physical quantities, man has arbitrarily chosen units which are both convenient and independent of the relevant physical phenomena. For reasons of convenience, the units are of two kinds : fundamental and derived. It follows that the correct mathematical expression of a physical phenomenon should be independent of the system of units adopted. Relation (3.2) should possess a particular feature which will make it independent of the units and should remain valid even though the units are changed. According to some authors [1.4] [4.1], this condition is expressed by the principle of the invariance of physical laws, which can be stated thus : *laws governing physical phenomena are invariant, irrespective of the change in the units.*

The condition required by the particular characteristic considered above is that relation (3.2) be "homogeneous" with respect to any fundamental physical quantity. Therefore, "homogeneity" will be called that *particular feature of a physical relation whereby it contains the same fundamental physical quantities in both its members* [1.44]. Whenever a homogeneous equation is involved, the change in the measuring units will only require the relation to be multiplied by a numerical factor, with no other change whatsoever. Thus, a simple rule can be established : *all terms of a physical relation should be dimensionally homogeneous.* In this book, homogeneity should be taken to mean "dimensional homogeneity".

One can never overemphasize the importance of this rule. Its applications to theoretical developments has yielded some results. This rule is now accepted as a metaphysical principle governing any physical theory. Hence it follows that the relation of definitions of secondary physical quantities should comply with the dimensional homogeneity condition ; likewise, it further follows that an empirical relation disregarding this condition should be considered to be the expression of an incorrectly known physical law [1.32].

In some physical relations the homogeneity is concealed ; such are those relations containing dimensional constants. In adapted relations, homogeneity is conditioned by those units which are used for the measurement of dimensional constants. Generally, in the empirical equations, homogeneity does not appear.

4.1 p. 42.

5 Dimensional Equations

5.1 Mathematical Definition of Dimension

The meaning of dimension is still highly controversial. The concept of dimension is basic and of mathematical origin. In this field, by dimension we mean a number which is one of the determinant elements of a mathematical quantity (a real, complex or hypercomplex number, a vector, a tensor, a matrix). A real number has therefore only one dimension, whilst a complex number has two, a hypercomplex number of order n has n dimensions, a vector has three, a second-order tensor (in the three dimensional space) has nine, and a matrix of m-lines and n-columns has mn dimensions. This definition of the term is the original one.

In geometry, dimension is defined as the real quantity, determining — alone or together with other physical quantities — the position of a point in a continuous multiplicity (line, surface, three-dimensional space etc.); it is also the determinant for the extension of a measurable geometrical figure.

5.2 Fundamental Theorem of Dimensional Analysis

Any relation expressing a physical phenomenon implies primary and secondary physical quantities. Primary quantities are chosen for their intrinsic qualities, for each field of science. Secondary quantities are expressed, as a function of the primary ones, through a variety of relations of definition.

The validity of a homogeneous physical equation should not be affected by a change in the units. A change in the units can be accounted for at any time. In doing so, we have to use fundamental and derived physical quantities. Fundamental quantities are chosen at random; the derived ones are expressed as a function of the fundamental ones, by means of the relations of definition used for the secondary physical quantities. In order to maintain consistency between physical quantities and physical relations, the following rules should be borne in mind when defining the secondary physical quantities :

— *there is no advantage in having numerical coefficients in the relations of definitions* ;

— *there is an advantage in having the relations of definition resulting in physical quantities which could lead to the establishment of homogeneous relations.*

Of course, nothing stops us from defining a physical quantity P, derived from, say, the physical quantities A, B, C, . . ., by means of a transcedental equation containing logarithms, exponential functions, circular functions etc. However, in such a case, no homogeneous equations could be achieved. Thus, for instance, a physical quantity defined by the relation $m \log v$, where m is the mass of a body while v is its velocity of motion, cannot be taken into consideration for two reasons, namely it implies a numerical coefficient ; and it does not satisfy the homogeneity condition.

Owing to the correlation existing between the primary and the secondary physical quantities, on one hand, and the fundamental and derived ones on the other, the homogeneity condition, which — as known—refers to fundamental and derived physical quantities, is adequately reflected in the primary and secondary physical quantities ; hence, the two classes of physical quantities should be defined by relations of like form. The homogeneity of physical relations can only be secured if the derived physical quantities are in a certain correlation with the fundamental physical quantities. This correlation is stated by the fundamental theorem of dimensional analysis.

In order to formulate this theorem, the following should be known :

— *the ratio of the measures of two states of a physical quantity is a number which exclusively depends on the two states of the considered physical quantity*. Thus, for instance, the ratio of the lengths of measures 5 m and 9 m respectively is the number 5/9. This ratio is independent of the selected system of units. In the above example, the ratio would have the same value if the lengths were of 500 cm and 900 cm respectively ;

— *any change in the fundamental units will result in multiplying the measure of the physical quantities by a variety of factors.* For example : when the unit "metre" is changed for the unit "centimetre", the number of the physical quantity is increased hundredfold.

Let us consider any physical quantity, say G, in the field of mechanics. In this field units corresponding to the fundamental physical quantities of length L, mass M and time T are currently used. We can therefore write

$$g = f(l, m, t). \tag{5.1}$$

This relation is correct only when the equality of the two members is observed for the measure of the physical quantity G and of the fundamental physical quantities L, M, T, as well as for the units. The symbol g will therefore represent both the measure of the physical quantity G and the derived unit ; likewise, symbols l, m, t will represent both the measures of the fundamental physical quantities and the fundamental units.

Let us now consider two states, G_1 and G_2 of the physical quantity G. The measurement of the physical quantity G in the two states yields

$$g_1 = f(l_1, m_1, t_1) \tag{5.2}$$

for the first state and

$$g_2 = f(l_2, m_2, t_2) \tag{5.3}$$

for the second. Now let us change the fundamental units and measure again the two states of the physical quantity G with these new units. The result will be :

$$g_1' = f(\lambda l_1, \mu m_1, \tau t_1) \tag{5.4}$$

and

$$g_2' = f(\lambda l_2, \mu m_2, \tau t_2) \tag{5.5}$$

respectively, λ, μ and τ being the factors attached to the change of the fundamental units ; they have arbitrary values.

Based on what was said above, we can write

$$\frac{g_1'}{g_2'} = \frac{f(\lambda l_1, \mu m_1, \tau t_1)}{f(\lambda l_2, \mu m_2, \tau t_2)} = \text{const.} \tag{5.6}$$

The symbols l, m, t in (5.6) should be considered as symbols of the fundamental units. Now, relation (5.6) is derived with respect to the arbitrary factors λ, μ and τ; for the sake of simplicity, we write only the derivative with respect to λ

$$\frac{\partial f(\lambda l_1, \mu m_1, \tau t_1)}{\partial \lambda f(\lambda l_2, \mu m_2, \tau t_2)} = 0; \tag{5.7}$$

after computation, we find :

$$\frac{\frac{\partial}{\partial \lambda} f(\lambda l_1, \mu m_1, \tau t_1)}{f(\lambda l_1, \mu m_1, \tau t_1)} = \frac{\frac{\partial}{\partial \lambda} f(\lambda l_2, \mu m_2, \tau t_2)}{f(\lambda l_2, \mu m_2, \tau t_2)}. \tag{5.8}$$

It follows from (5.8) that indexes can be neglected and hence :

$$\frac{\frac{\partial}{\partial \lambda} f(\lambda l, \mu m, \tau t)}{f(\lambda l, \mu m, \tau t)} = \text{const.} = l_g. \tag{5.9}$$

Evidently

$$\frac{\partial}{\partial \lambda} f(\lambda l, \mu m, \tau t) = \frac{\partial f(\lambda l, \mu m, \tau t)}{\partial \lambda l} \cdot \frac{\partial \lambda l}{\partial \lambda} \tag{5.10}$$

and

$$\frac{\partial \lambda l}{\partial \lambda} = l; \tag{5.11}$$

it follows that :

$$\frac{l \, \frac{f(\lambda l, \mu m, \tau t)}{\partial \lambda l}}{f(\lambda l, \mu m, \tau t)} = l_g. \tag{5.12}$$

This relation should be independent of λ; when $\lambda = 1$, the result is

$$\frac{\partial f(l, \mu m, \tau t)}{f(l, \mu m, \tau t)} = l_g \frac{\partial l}{l}. \tag{5.13}$$

Integrating relation (5.13) results in:

$$\log f(l,\ \mu m,\ \tau t) = l_g \log l + \log C_1 \tag{5.14}$$

where C_1 is an integration constant; this relation leads to

$$f(l,\ \mu m,\ \tau t) = C_1 l^{l_g}. \tag{5.15}$$

The complete calculations of the arbitrary factors μ and τ lead to similar results; therefore, we can write the final relation:

$$f(l,\ m,\ t) = C l^{l_g}\ m^{m_g}\ t^{t_g}. \tag{5.16}$$

On the basis of the above demonstration, from (5.1) and (5.16), the final relation is

$$g = C l^{l_g}\ m^{m_g}\ t^{t_g}. \tag{5.17}$$

This result can be stated as follows: *the derived unit should be expressed as a function of the fundamental units through a monomial relation.* Relation (5.17) is the dimensional equation (the measure formula) of unit g, derived from the fundamental units. It is called, in brief, the dimension of the derived unit. Constants l_g, m_g and t_g are called dimensional subscripts.

In the demonstration above, relation (5.1) was considered to be continuous and derivable. Birkhoff arrives at the same result, although he uses only the continuity, and not the derivable quality of relation (5.1); a similar demonstration is given in [1.18].

Referring to this demonstration, known from the literature [1.11], [1.36], Charcosset stated not very long ago [1.26] [5.1] that by repeating the calculations for the other factors implied in the change of units μ and τ, the result arrived at is not the one given by relation (5.17); instead, calculations lead to the relation

$$C_1 l^{l_g} = C_2 m^{m_g} = C_3 t^{t_g}. \tag{5.18}$$

5.1 Charcosset, C., *Démonstration du Théorème* II in: [1.26], Appendix 1, p. 205.

As a result, Charcosset suggested a new demonstration for the fundamental theorem of dimensional analysis. He considered

$$\frac{g_1}{g_2} \equiv \frac{g_1'}{g_2'} \tag{5.19}$$

which, as we can appreciate from above, is evident. Now, if relations (5.2)—(5.5) are introduced in relation (5.19):

$$\frac{f(\lambda l_1,\ \mu m_1,\ \tau t_1)}{f(l_1,\ m_1,\ t_1)} \equiv \frac{f(\lambda l_2,\ \mu m_2,\ \tau t_2)}{f(l_2,\ m_2,\ t_2)}\ ; \tag{5.20}$$

and from what was said above:

$$\frac{f(\lambda l,\ \mu m,\ \tau t)}{f(l,\ m,\ t)} \equiv k(\lambda,\ \mu,\ \tau). \tag{5.21}$$

Relation (5.21) can be put in the form

$$f(\lambda l,\ \mu m,\ \tau t) \equiv f(l,\ m,\ t)k(\lambda,\ \mu,\ \tau)\ ; \tag{5.22}$$

it is clear that permuting the symbols l, m, t and λ, μ, τ, respectively, will in no way affect the first member of relation (5.22); it follows that the correct rendering will be

$$f(l,\ m,\ t)\ k(\lambda,\ \mu,\ \tau) \equiv f(\lambda,\ \mu,\ \tau)\ k(l,\ m,\ t). \tag{5.23}$$

Relation (5.23) might be also put is the form

$$\frac{f}{k}\ (l,\ m,\ t) \equiv \frac{f}{k}\ (\lambda,\ \mu,\ \tau)\ ; \tag{5.24}$$

now, assuming that variables l, m and t can be independently varied, then

$$\frac{f}{k} = \text{const.} = X. \tag{5.25}$$

As a consequence, relation (5.22) becomes

$$f(\lambda l,\ \mu m,\ \tau t) \equiv \frac{1}{X} f(l,\ m,\ t) f(\lambda,\ \mu,\ \tau). \tag{5.26}$$

It is demonstrated that constant X can be assumed to be equal to unity [1.26] [5.2]. In such a case, relation (5.26) becomes

$$f(\lambda l,\ \mu m,\ \tau t) \equiv f(l,\ m,\ t) f(\lambda,\ \mu,\ \tau). \tag{5.27}$$

The following question arises : what is the form of function f from relation (5.27)? To answer this question one should bear in mind that in the above demonstration, fundamental units were used ; these can be chosen at random. This hypothesis is mathematically translated by the freedom to independently change the numbers λ, μ and τ.

Now relation (5.27) is derived. For the sake of simplification, only the derivation with respect to l will be considered, that is

$$\frac{\partial f}{\partial u}(\lambda l,\ \mu m,\ \tau t)\,\frac{\mathrm{d}u}{\mathrm{d}l} = \frac{\partial f}{\partial u}(l,\ m,\ t)\,\frac{\mathrm{d}u}{\mathrm{d}l} f(\lambda,\ \mu,\ \tau) \tag{5.28}$$

where

$$\lambda l = u \text{ and } l = u \tag{5.29}$$

in the first and second members respectively ; it is evident that

$$\frac{\mathrm{d}u}{\mathrm{d}l} = \lambda \text{ and } \frac{\mathrm{d}u}{\mathrm{d}l} = 1. \tag{5.30}$$

Relation (5.28) is multiplied by l ; bearing in mind relations (5.29) and (5.30), relation (5.28) becomes

$$\left(u\frac{\partial f}{\partial u}\right)(\lambda l,\ \mu m,\ \tau t) = \left(u\frac{\partial f}{\partial u}\right)(l,\ m,\ t) f(\lambda,\ \mu,\ \tau). \tag{5.31}$$

[5.2] p. 209.

Considering that relation (5.27) can be written in the form

$$f(\lambda, \mu, \tau) = \frac{f(\lambda l, \mu m, \tau t)}{f(l, m, t)}, \tag{5.32}$$

relation (5.31) will take the form

$$\left(\frac{u}{f} \cdot \frac{\partial f}{\partial u}\right)(\lambda l, \mu m, \tau t) = \left(\frac{u}{f} \cdot \frac{\partial f}{\partial u}\right)(l, m, t). \tag{5.33}$$

This relation becomes

$$\frac{u}{f} \cdot \frac{\partial f}{\partial u} = \text{const.} = l_g, \tag{5.34}$$

where l_g is a constant independent of l, m and t, as well as of λ, μ and τ. From (5.34), it follows by integration:

$$f = C_1 u^{l_g}, \tag{5.35}$$

where C_1 is an integration constant. Similar computations carried out for the variables m and t lead to similar results. Thus, eventually, one can write

$$\mu m = v \text{ and } m = v, \tag{5.36}$$

and

$$\tau t = w \text{ and } t = w, \tag{5.37}$$

as well as the complete relation

$$f = C u^{l_g} v^{m_g} w^{t_g}. \tag{5.38}$$

The above relation, along with the relations found for variables u, v and w, leads to

$$f = C l^{l_g} m^{m_g} t^{t_g}. \tag{5.39}$$

It follows directly from relations (5.1) and (5.39) that

$$g = Cl^{l_g} m^{m_g} t^{t_g} \tag{5.17}$$

that is, a result previously obtained. Variables l, m and t can take arbitrary values. If $l = m = t = 1$, the result is $g = 1$; and hence the integration constant will be given by $C = 1$. Consequently, the final form of relation (5.17) will be

$$g = l^{l_g} m^{m_g} t^{t_g}. \tag{5.40}$$

5.3 Dimensional Equation of the Physical Quantity

Fourier was the first to apply the dimensional concept to physical quantities. This made Palacios say that Fourier should be considered to be the promoter of the dimensional analysis [1.32] [5.3]. In his work, published in 1822 under the title "*Théorie analytique de la chaleur*" (Analytical Theory of Heat), Fourier says that "one should bear in mind that every known — or unknown — physical quantity has its own dimension and that the terms of an equation cannot be compared against one another unless they have the same dimensional subscript". The reason why Fourier introduced this concept of "dimension" to the analytical theory of heat is to make the definitions "more steady" and to check if the calculations were correct. The numbers representing surfaces or volumes take the dimension two in the first case, and three in the second. Angles, trigonometric functions and logarithms are absolute numbers which do not change with the units; therefore, their dimension is zero.

Fifty years after Fourier, in 1871, Maxwell offered a new notation for the dimensions of physical quantities: he writes the dimensions of a physical quantity in the form of a product of powers and, in order to underline the fact that such products refer to dimensions, he introduces square brackets.

Let us consider, for instance, the fundamental law of dynamics. This relation should maintain its mathematical form even when, in a given system of units, the physical quantity of the entities or the

---

[5.3] p. VII.

physical quantity of the units are varied. Equation (2.9) should therefore keep its form, so :

— either the units are made φ—, μ— or α— times smaller respectively ;

— or, the corresponding physical quantities are made φ—, μ—, or α— times greater respectively. In both cases the result is :

$$\varphi F = \mu m \cdot \alpha a, \tag{5.41}$$

from which

$$\varphi = \mu \alpha. \tag{5.42}$$

Relation (5.42) expresses the unit of force as a function of the units of mass and acceleration ; this relation is of "the same form" as relation (2.9) expressing the physical law. Symbols φ, μ, α, represent ratios of units (or physical quantities) and are numbers ; therefore, the rules of the algebraic calculation can be applied. Thus, for instance, from equation (5.42) and the equation

$$\alpha = \lambda \tau^{-2} \tag{5.43}$$

which was derived from the acceleration definition relation as shown above, it follows directly that

$$[F] = \mathrm{LMT}^{-2}, \tag{5.44}$$

if the corresponding physical quantities are taken into consideration. Relation (5.44) is the dimensional equation for the physical quantity of force.

The demonstration given for this particular case is generally valid. It follows that the dimension of a physical quantity is expressed through the product of literal symbols assigned to fundamental physical quantities raised to the powers specified by the dimensional exponents ; it is represented by the literal symbol of the relevant physical quantity put within square brackets. The dimensional equation of the physical quantity A, for example, considered to pertain to mechanics, will therefore have the form

$$[A] = \mathrm{L}^{l_a}\mathrm{M}^{m_a}\mathrm{T}^{t_a} \tag{5.45}$$

in the system of units having length L, mass M and time T as fundamental physical quantities.

From what has been said above, an important conclusion can be drawn : the *sine qua non* condition to be born in mind when writing the relation of definition for a secondary physical quantity is that the dimensional equation derived from that relation should be of the product-of-powers type (5.45) ; only in this case, can the homogeneity of physical equations be secured.

The dimensional equation of a physical quantity can be thus considered as an envelope enclosing a relation of definition or a physical law. The dimensional equation of any physical quantity follows directly from the relation of definition written for the corresponding physical quantity ; the code of dimensions is not difficult to establish. It was demonstrated that the dimensional equation is always of the form of a product of powers. Thus, for instance, the dimension of the xy product is written : $[xy] = XY$; the dimension of the $\mathrm{d}y/\mathrm{d}x$ derivative is written : $[\mathrm{d}y/\mathrm{d}x] = X^{-1}Y$; the dimension of the integral $\int y\mathrm{d}x$ is written : $\left[\int y\mathrm{d}x\right] = XY$ etc. The dimension of a surface (area) is $[S] = \mathrm{L}^2$; that of a velocity is $[v] = \mathrm{LT}^{-1}$ etc. To those who think that dimensions have an intrinsic significance, the above relations might appear as paradoxical. However, it can be shown that any attempt to establish the true or real dimension of a physical quantity is of no effect. Furthermore, it is misleading to seek in dimension any mysterious quality which could reveal something of the very inmost nature of things. Dimension is a result of arbitrarily accepted elements and varies as these vary ; therefore it is only a specific characteristic of the physical quantity. Planck said the desire to know the real dimension of an entity is as significant as the desire to know the true name of an object [1.26].

It may be that, in a given system of units, two physical quantities have the same dimensional equation. But it does not necessarily follow that they are physical quantities of the same nature or that they share in common the same physical character. They are only dimensionally equivalent, i.e. the ratio of their numerical values in the relevant system of units is independent of the physical quantity of the fundamental units. This dimensional identity is not necessarily maintained in other systems of units.

Appendix 2 presents the dimensional equations, written in IS units, for the physical quantities typical of the fields to which most of the applications of dimensional analysis discussed in this work pertain.

*

Speaking of the dimensional equation, Huntley suggested separating three components for the fundamental physical quantity of length L, these components following the directions of a spatial system of Cartesian co-ordinates [1.32] [5.4]. He further suggested taking into consideration the distinction made by the theory of general relativity, between the inertial mass m_i and the gravitational mass m_g. As a consequence, a new system of units with six fundamental physical quantities could be adopted.

Considering mass as a fundamental physical quantity, it can be demonstrated that adopting the distinction between inertial mass and gravitational mass makes no difference whatsoever. Therefore, the dimensional equation of physical quantity A will have the form

$$[A] = \mathrm{L}_x^{la}\,\mathrm{L}_y^{o}\,\mathrm{L}_z^{o}\,\mathrm{M}^{ma}\,\mathrm{T}^{ta}. \tag{5.46}$$

The distinction suggested by Huntley obviously increases the number of equations worked out between the dimensional and the physical quantity exponents to express the homogeneity condition. This is an important advantage since the solution will be more accurate. Huntley called this procedure "the method of fundamental dimensions components". Palacios called it "the method of spatial dimensions separation".

6 Uses of Dimensional Equations

6.1 Homogeneity of Physical Relations

Dimensional equations of physical quantities provide an easy check on the correctness of physical relations, insofar as their homogeneity is concerned. In mechanics, for instance, any relation which does not show a triple homogeneity — with respect to length, mass

[5.4] p. 59.

and time — is categorically wrong. However, the reciprocal is not true: it does not necessarily follow that a homogeneous relation is also physically valid. This check is a simple one and is always recommended; to this end the dimensional equations of physical quantities used to describe the phenomenon should be available.

In mathematics, a square table of numbers is called a matrix; when the numbers of the matrix represent dimensional exponents, the table is called a dimensional matrix.

Let us consider, for example, the relation

$$Fv = \rho a, \tag{6.1}$$

where the symbols represent force F, velocity v, density ρ and the acceleration a respectively. The dimensional equations of these physical quantities are derived from the following dimensional matrix

	F	v	ρ	a
L	1	1	−3	1
M	1	0	1	0
T	−2	−1	0	−2

(6.2)

It can easily be seen that relation (6.1) is not valid since the dimension of the first member differs from that of the second.

A check on the homogeneity of physical relations will not disclose any sign error or the omission of a term; also it cannot lead to the value of a numerical factor.

6.2 Value of the Physical Quantity when the Units of Measurement are Changed

At the present stage of scientific development, there are situations when a change-over from a system of units to another is required. The reasons for such a conversion may be various: the existence of several systems of units, the arbitrary use of the systems of units, correlation with the specific character of a practical field, tradition, educational purposes etc. Let us consider an acceleration of 100 cm/s² which has to be expressed in km/h². As known:

$$1 \text{ cm} = 10^{-5} \text{ km}, \ 1 \text{ s} = 3.6^{-1} \cdot 10^{-3} \text{ h}. \tag{6.3}$$

It follows therefore that:

$$100 \text{ cm/s}^2 = \frac{100 \cdot 10^{-5} \text{ km}}{(3.6^{-1} \cdot 10^{-3} \text{ h})^2} = 12\,960 \text{ km/h}^2. \tag{6.4}$$

Let us take another example: it is necessary to express in grammes the value of the astronomical units for mass, m_a, in the astronomical system of units where the units of length and time are the kilometre and the second respectively. The law of universal gravitation is expressed through relation (2.10) where the universal constant $k' = 6.7 \cdot 10^{-8} \dfrac{\text{cm}^3}{\text{gs}^2}$, in the CGS system of units. By definition, in the astronomical system of units, $k'=1$. The correlation between the values of the physical quantities of units is expressed through relations:

$$1 \text{ cm} = 10^{-5} \text{ km}, \ 1 \text{ g} = x \cdot \text{m}_a. \tag{6.5}$$

With these values, it follows that:

$$k' = 6.7 \cdot 10^{-8} \frac{\text{cm}^3}{\text{gs}^2} = 6.7 \cdot 10^{-8} \frac{(10^{-5} \text{ km})^3}{x \text{m}_a \cdot \text{s}^2} = 1 \frac{\text{km}^3}{\text{m}_a \cdot \text{s}^2}; \tag{6.6}$$

from (6.6), we obtain:

$$x = 6.7 \cdot 10^{-8} \cdot 10^{-15} = 6.7 \cdot 10^{-23}. \tag{6.7}$$

The first relation from (6.5) and relation (6.7) produce

$$1 \text{ g} = 6.7 \cdot 10^{-23} \text{ m}_a \tag{6.8}$$

or in another way, the value of the mass unit in the astronomical system of units is

$$1 \text{ m}_a = \frac{1}{6.7 \cdot 10^{-23}} = 1{,}492 \cdot 10^{22}\text{g}. \tag{6.9}$$

6.3 Significance of an Element of a Relation

In physical relations with several terms, the meaning of some symbols may be found by means of dimensional equations. Take for instance relation

$$S = a + bt + ct^2, \tag{6.10}$$

where S is the length and t is the time ; then a is a length too, while b should have the nature of a velocity and c, of an acceleration.

Dimensional equations can demonstrate that the physical quantity

$$L = mv^2, \tag{6.11}$$

where m is the mass and v is the velocity, is of the nature of either a momentum or mechanical work since :

$$[mv^2] = \text{ML}^2\text{T}^{-2} = \text{LMT}^{-2} \cdot \text{L} = [F]\,[l]. \tag{6.12}$$

Similarly, the physical significance of the universal constants can be found. Thus, in Newton's law, (2.10), the constant

$$k' = \frac{Fl^2}{mm'} \tag{6.13}$$

has the dimension :

$$[k'] = \frac{\text{LMT}^{-2} \cdot \text{L}^2}{\text{M}^2} = \text{L}^3\text{M}^{-1}\text{T}^{-2}. \tag{6.14}$$

The dimensional equation of the proportionality factor σ from the relation experimentally established by Stefan for the energy w radiated by a body at temperature θ, results from the relation

$$[\sigma] = \frac{[w]}{[\theta]}. \tag{6.15}$$

Based on the unit system with four fundamental physical quantities, the dimensional equations of the physical quantities w and θ follow from the following dimensional matrix:

	w	θ
L	-1	0
M	1	0
T	-2	0
Θ	0	1

(6.16)

Relation (6.15) produces

$$[\sigma] = \mathrm{L}^{-1}\mathrm{M}\mathrm{T}^{-2}\Theta^{-4}. \tag{6.17}$$

The validity of Stefan's relation was further confirmed by Boltzmann who showed that this law is a consequence of the thermodynamics principles, as well as of the fact that radiation exerts a pressure. By applying the quantum theory, it was shown that

$$\sigma = K \frac{k^4}{v^3 h^3}, \tag{6.18}$$

where the symbols represent Boltzmann's constant k, the velocity of light v and Planck's constant h; factor K has a numerical character this time. For the physical quantities in (6.18) the dimensional matrix is:

	k	v	h
L	2	1	2
M	1	0	1
T	-2	-1	-1
Θ	-1	0	0

(6.19)

It can be seen that relations (6.18) and (6.19) lead to relation (6.17).

6.4 Conversion Relation for the Units of Measurement

Dimensional equations unequivocally describe how the measures of physical quantities change as a result of a change in the units. They may therefore serve whenever a quick conversion is requested from one system of units into another system of units. The changing procedure could be expressed through a simple formula.

Let us consider any physical quantity, say X, from mechanics. The dimensional equation of this physical quantity will be

$$[X] = \mathrm{L}^{l_x}\,\mathrm{M}^{m_x}\mathrm{T}^{t_x}, \tag{6.20}$$

where the length L, the mass M and the time T are fundamental physical quantities. In an initial system of units, the measure of X is x. We shall adopt a new system of units, correlated with the first one as follows:

— the old unit of length $= \lambda$ new units of length;
— the old unit of mass $= \mu$ new units of mass;
— the old unit of time $= \tau$ new units of time.

In the new system of units, the measure of the physical quantity considered is $\bar{x}$.

The relation expressing the change in the units is the solution to the problem and will therefore refer to the relationship between measures x and $\bar{x}$. We shall denote the old units of length, mass and time by OL, OM, and OT respectively; we shall further denote the new units by NL, NM and NT respectively. Now, from what was said above, we can write:

$$\mathrm{OL} = \lambda\mathrm{NL},\ \mathrm{OM} = \mu\mathrm{NM},\ \mathrm{OT} = \tau\mathrm{NT}. \tag{6.21}$$

Bearing in mind relation (6.20), the measure of physical quantity X will be written:

$$x[(\mathrm{OL})^{l_x}\ (\mathrm{OM})^{m_x}\ (\mathrm{OT})^{t_x}] = x[(\lambda\mathrm{NL})^{l_x}\ (\mu\mathrm{NM})^{m_x}\ (\tau\mathrm{NT})^{t_x}] =$$
$$= \bar{x}[(\mathrm{NL})^{l_x}\ (\mathrm{NM})^{m_x}\ (\mathrm{NT})^{t_x}]; \tag{6.22}$$

from which it directly follows that

$$\bar{x} = x\lambda^{l_x}\ \mu^{m_x}\ \tau^{t_x}. \tag{6.23}$$

Relation (6.23) represents the algebraic conversion formula for the measure of a physical quantity when a change in the measuring units occurs. As can be seen from the above relation, it follows that : *in order to convert the measure of a physical quantity, as expressed by given fundamental units, into the measure of the same physical quantity as expressed by other fundamental units, the original measure shall be multiplied by the product of the conversion factors of the fundamental units, each factor being raised to the power of its corresponding dimensional exponent* [1.4].

For the example of the unit conversion given above (Section I.6.2) and on the basis of what was said above, we may write

$$x = 100 \text{ cm/s}^2 \qquad l_x = 1$$
$$\lambda = 10^{-5} \qquad t_x = -2. \tag{6.24}$$
$$\tau = 3.6^{-1} \cdot 10^{-3}$$

Relation (6.23) leads to

$$\bar{x} = 100(10^{-5})^1 \cdot (3.6^{-1} \cdot 10^{-3})^{-2} = 12{,}960 \text{ km/h}^2 \tag{6.25}$$

which is identical with what was directly obtained from (6.4).

6.5 Conversion Relations for Fundamental Physical Quantities

The dimensional equation of a physical quantity, according to the arbitrarily selected fundamental physical quantities, depends upon the nature and number of the fundamental units; this is to say that such an equation changes with the change in the nature and number of the fundamental units. Any change of fundamental physical quantities leads to certain relations among the dimensional exponents of the old system of units and those of the new one; such relations are expressed through simple formulae too.

Let us consider any physical quantity X and a system of units having as physical quantities $M_1, \ldots, M_m$. According to relation (5.45), the dimension of physical quantity X will be written

$$[X] = M_1^{a_1} \ldots M_m^{a_m}. \tag{6.26}$$

Now, we choose a new system of units having as fundamental physical quantities $M'_1, \ldots, M'_m$; the dimensional equation of the physical quantity X will be

$$[X] = (M'_1)^{a'_1} \ldots (M'_m)^{a'_m}. \tag{6.27}$$

The old fundamental physical quantities have the following dimensional equations as functions of the new fundamental physical quantities:

$$\begin{aligned} [M_1] &= (M'_1)^{\alpha_{11}} \ldots (M'_m)^{\alpha_{1m}} \\ &\cdot \; \cdot \; \cdot \; \cdot \; \cdot \; \cdot \; \cdot \; \cdot \; \cdot \; \cdot \; \cdot \\ [M_m] &= (M'_1)^{\alpha_{m1}} \ldots (M'_m)^{\alpha_{mm}}. \end{aligned} \tag{6.28}$$

Introducing equations (6.28) in relation (6.26) results in

$$[X] = ((M'_1)^{\alpha_{11}} \ldots (M'_m)^{\alpha_{1m}})^{a_1} \ldots ((M'_m)^{\alpha_{m1}} \ldots (M'_m)^{\alpha_{mm}})^{a_m}. \tag{6.29}$$

By comparison, from (6.27) and (6.29) we obtain

$$\begin{aligned} a'_1 &= \alpha_{11} \cdot a_1 + \ldots + \alpha_{m1} \cdot a_m \\ &\cdot \; \cdot \; \cdot \; \cdot \; \cdot \; \cdot \; \cdot \; \cdot \; \cdot \; \cdot \; \cdot \; \cdot \; \cdot \\ a'_m &= \alpha_{1m} \cdot a_1 + \ldots + \alpha_{mm} \cdot a_m \end{aligned} \tag{6.30}$$

which are the relationships among the dimensional exponents.

Let us consider, for example, a system of units having three fundamental physical quantities L, M and T, along with the astronomical system of units having two fundamental physical quantities: L and T. In this latter system, the dimensional equation of the physical quantity for mass is derived from relation (2.9) and is of the form

$$[m] = L^3T^{-2}; \tag{6.31}$$

the fundamental quantities L and T are common to the two systems of units. For a physical quantity X we therefore can write:

$$[X] = L^{l_x}M^{m_x}T^{t_x} = L^{l_x+3m_x}\,T^{t_x-2m_x} \tag{6.32}$$

where due consideration was paid to relation (6.31). Consequently, the values of the dimensional exponents l'_x and t'_x in the system of units with two fundamental physical quantities will be :

$$l'_x = l_x + 3m_x$$
$$t'_x = t_x - 2m_x. \qquad (6.33)$$

For example, the dimensional equation for force will be

$$[F] = \mathrm{LMT}^{-2} \qquad (6.34)$$

in the system of units with three fundamental physical quantities, while in the system with two fundamental physical quantities it will be

$$[F] = \mathrm{L}^4\mathrm{T}^{-4}. \qquad (6.35)$$

It is demonstrated that in the case of a conversion from the system of units with four fundamental physical quantities to that with two fundamental physical quantities, relation (6.33) becomes :

$$l'_x = l_x + 3m_x + 4f_x,\ t'_x = t_x - 2m_x - 4f_x, \qquad (6.36)$$

while if the new system of units has three fundamental physical quantities, the following conversion relations are valid :

$$l'_x = l_x + f_x,\ m'_x = m_x + f_x,\ t'_x = t_x - 2f_x. \qquad (6.37)$$

In the latter case, the dimension of mass is, for instance

$$[m] = \mathrm{M} \qquad (6.38)$$

in the system of units with four fundamental physical quantities, and

$$[m] = \mathrm{L}^{-1}\mathrm{FT}^2 \qquad (6.39)$$

when written in the system with three fundamentally physical quantities.

7 Conditions of Dimensional Homogeneity

7.1 Degrees of Homogeneity

Let us consider relation (3.10), which is the most general form of the physical relation written for measuring units:

$$f(l, m, t) = \varphi(l, m, t). \tag{7.1}$$

By changing the fundamental units, we may write

$$f(l', m', t') = \varphi(l', m', l'). \tag{7.2}$$

Functions f and φ should be of the like form as in relation (7.1); this is only possible if certain conditions are secured.

We shall admit that the change in the units could be expressed by the affinity relations:

$$l' = \lambda l, \; m' = \mu m, \; t' = \tau t. \tag{7.3}$$

Substituting relations (7.3) in (7.2) results in

$$f(\lambda l, \mu m, \tau t) = \varphi(\lambda l, \mu m, \tau t). \tag{7.4}$$

Bearing in mind relation (5.40), we can write for the original units:

$$f(l, m, t) = l^{\alpha} m^{\beta} t^{\gamma}, \; \varphi(l, m, t) = l \quad m^{\beta_1} t^{\gamma_1} \tag{7.5}$$

and for the new units:

$$f(l', m', t') = l'^{\alpha} m'^{\beta} t'^{\gamma}, \; \varphi(l', m', t') = l'^{\alpha_1} m'^{\beta_1} t'^{\gamma_1}. \tag{7.6}$$

Now we substitute relations (7.3) in (7.6); bearing in mind relations (7.5) and it follows that

$$(\lambda l, \mu m, \tau t) = \lambda^{\alpha} \mu^{\beta} \tau^{\gamma} f(l, m, t), \; \varphi(\lambda l, \mu m, \tau t) = \lambda^{\alpha_1} \mu^{\beta_1} \tau^{\gamma_1} \varphi(l, m, t). \tag{7.7}$$

Relations (7.7) show that functions f and φ are homogeneous. Relations (7.4) and relations (7.7) lead to:

$$\lambda^{\alpha} \mu^{\beta} \tau^{\gamma} f(l, m, t) = \lambda^{\alpha_1} \mu^{\beta_1} \tau^{\gamma_1} \varphi(l, m, t) \tag{7.8}$$

which is identical with (7.1) only if:

$$\alpha = \alpha_1 ; \beta = \beta_1 ; \gamma = \gamma_1. \tag{7.9}$$

The result thus obtained represents the condition of homogeneity of the physical relations and can be stated as follows: *the two members of a relation expressing a physical law must have equal degrees of homogeneity with respect to each of the fundamental units.*

7.2 Conversion Factors for the Units of Measurement

Let us consider that in the case of a physical phenomenon taken from, say, mechanics, the physical quantity P is a variable dependent upon $A, B, C, \ldots, M, N, \ldots$ which are independent variables. The relation of physical quantities can be written in the form already known:

$$P = f(A, B, C, \ldots M, N, \ldots). \tag{3.3}$$

The relation for the corresponding measures of physical quantities $a, b, c, \ldots, m, n, \ldots$, which will be obtained in a certain system of units, will therefore have a similar form

$$p = f(a, b, c, \ldots, m, n, \ldots), \tag{7.10}$$

where symbol f is to be considered as an operator, applied to the independent variables to produce the measure of the dependent variable. Assuming that the system of units is changed, the variables considered above will take the new values $\bar{a}, \bar{b}, \bar{c}, \ldots, \bar{m}, \bar{n}, \bar{p}, \ldots$; based on the definition accepted for the homogeneity of physical relations (Section I.3), the equation written for the new values of the variables will be dimensionally homogeneous if, and only if

$$\bar{p} = f(\bar{a}, \bar{b}, \bar{c}, \ldots, \bar{m}, \bar{n}, \ldots) \tag{7.11}$$

where symbol f is the same operator as above. This mathematical condition expresses the idea that the relation is an "invariant" for all changes which may result from any conversion of units.

The dimensions of physical quantities expressed in (3.3) are represented by the following dimensional matrix:

$$\begin{array}{c|cccccc|} & A & B & C & \dots & P & \dots \\ \hline \mathrm{L} & l_a & l_b & l_c & \dots & l_p & \dots \\ \mathrm{M} & m_a & m_b & m_c & \dots & m_p & \dots \\ \mathrm{T} & t_a & t_b & t_c & \dots & t_p & \dots \\ \dots & \dots & \dots & \dots & \dots & \dots & \dots \end{array} \tag{7.12}$$

The change in the measure of a physical quantity subsequent to a conversion of units is defined with relation (6.23). Bearing in mind this relation, in our case we can write:

$$\begin{aligned} \bar{a} &= a\lambda^{l_a}\ \mu^{m_a}\ \tau^{t_a} = aK_a \\ \bar{b} &= b\lambda^{l_b}\ \mu^{m_b}\ \tau^{t_b} = bK_b \\ \bar{c} &= c\lambda^{l_c}\ \mu^{m_c}\ \tau^{t_c} = cK_c \\ &\dots\dots\dots\dots \end{aligned} \tag{7.13}$$

where λ, μ and τ are the conversion factors of the units. In relations (7.13), the new symbols K_a, K_b, K_c, ... whose meaning is evident, were introduced. Substituting relations (7.13) in relation (7.11) results in

$$K_p \cdot p = f(K_a \cdot a,\ K_b \cdot b,\ K_c \cdot c, \dots, K_m \cdot m, \dots); \tag{7.14}$$

now, bearing in mind relation (7.10), it follows from relation (7.14) that

$$K_p \cdot f(a, b, c, \dots, m, \dots) = f(K_a \cdot a, K_b \cdot b, K_c \cdot c, \dots, K_m \cdot m, \dots), \tag{7.15}$$

which could be stated as follows: *function $f(a, b, c, \dots, m, \dots)$ is dimensionally homogeneous if, and only if, relation (7.15) is an identity with respect to variables a, b, c, ..., m, ..., λ, μ, τ.*

For exemplification purposes, let us consider the physical phenomenon represented by the bending of a straight beam supported on two bearings placed at the beam ends; the beam is of length l, has an inertial momentum of its cross-section I and an elasticity modulus E. It is under the action of a normal force P which is concentrated on its middle section and generates the deflection s. Now, we can write the following relation between the dependent variable s and the independent variables l, I, E, P:

$$s = f(l,\ I,\ E,\ P). \tag{7.16}$$

According to relation (7.14), the condition for the dimensional homogeneity will be

$$K_s \cdot s = f(K_l \cdot l,\ K_I \cdot I,\ K_E \cdot E, K_P \cdot P), \tag{7.17}$$

which should be an identity with respect to variables λ, μ, τ. In such a case, the dimensional matrix will be:

	s	l	I	E	P
L	1	1	4	−1	1
T	0	0	0	1	1
M	0	0	0	−2	−2

(7.18)

According to (7.13), the relations for symbols K_s, K_l, K_I, K_E, K_P, are:

$$K_s = \lambda,\ K_l = \lambda,\ K_I = \lambda^4,\ K_E = \lambda^{-1}\mu\tau^{-2},\ K_P = \lambda\mu\tau^{-2}. \tag{7.19}$$

Based on these relations, we can see that relation (7.17) is an identity only if relation (7.16) is of the form

$$s = k\frac{l^3P}{IE} \tag{7.20}$$

where k is a pure number. It follows immediately that relation

$$\lambda_s = \frac{\lambda^3 l^3 \cdot \lambda\mu\tau^{-2}P}{\lambda^4 I \cdot \lambda^{-1}\mu\tau^{-2}E} \tag{7.21}$$

is an identity with respect to λ, μ and τ.

II. Restricted Dimensional Analysis

8 Form of Physical Relationships

8.1 The Π Theorem

Let us assume that all A, B, C, ..., M, N, P, ... variables which describe a definite physical phenomenon are known and that the state of this phenomenon or the variation of one physical quantity as a function of the change occurred in the other physical quantities can be described by a law. A relationship of the most general form can be written

$$f(A, B, C, \dots M, N, P, \dots) = 0 \tag{3.2}$$

which is the first step towards the study of any physical phenomenon. The next step refers to the form of relationship (3.2). The answer to this problem can be described as the object of the Π theorem. According to [1.26][8.1], Π theorem was formulated for the first time by Vaschy, in 1890, who published it in one of his works edited in 1892 [8.2]. This is the reason why Π theorem is often called Vaschy's theorem.

8.1 p. 205.

8.2 Vaschy, A., Sur les lois de similitude en physique ; *Annales Télégraphiques*, III, No. 11, 1892.

According to Vaschy, let us consider m^* physical quantities $A, B, C, \ldots, M, N, P, \ldots$ and a system of units of measurement having d^* fundamental physical quantities. We can further assume that the first d^* fundamental units $A, B, C, \ldots$ are referred to the fundamental units, while the remaining $m^* - d^*$ physical quantities, i.e. $M, N, P, \ldots$ are referred to the derived ones; for example: A could be a length, B a mass, C a measure of time, while the other physical quantities could be forces, velocities etc. Now, *if the m^* physical quantities are interconnected by a relation of the form (3.2), which subsists independently from the units of measurement chosen at random, this relation can be also written for at most $m^* - d^*$ parameters, i.e.*

$$\varphi(\Pi_m, \Pi_n, \ldots) = 0 \tag{8.1}$$

where parameters $\Pi_m, \Pi_n, \ldots$ are monomial functions of $A, B, C, \ldots \ldots, M, N, P, \ldots$.

In the particular case when $m^* = 5$ and $d^* = 3$, a theorem formulated by Carvallo [2.27] is re-established.

The theorem is demonstrated by taking into consideration the $m^* - d^*$ physical quantities $M, N, P, \ldots$ as referred to the derived units; according to this hypothesis, in the ratios:

$$\Pi_m = \frac{M}{A^{a_m} B^{b_m} C^{c_m}}, \quad \Pi_n = \frac{N}{A^{a_n} B^{b_n} C^{c_n}}, \ldots \tag{8.2}$$

exponents of a magnitude $a_m, b_m, c_m, \ldots, c_n, \ldots$ can be chosen so as to make the numerical values of ratios (8.2) independent of the arbitrary values of the fundamental units; if, for example, A, B, C and M represent a length, a mass, a time and a force respectively, then the value of the ratio

$$\Pi = \frac{M}{ABC^{-2}} \tag{8.3}$$

will be independent from the choice of the units; as a consequence, the relation (3.2) can be written in the form

$$f(A, B, C, \ldots, \Pi_m A^{a_m} B^{b_m} C^{c_m} \ldots, \Pi_n A^{a_n} B^{b_n} C^{c_n} \ldots, \ldots) = 0 \tag{8.4}$$

or a simpler one:

$$f(A, B, C, \ldots, \Pi_m, \Pi_n, \ldots) = 0. \tag{8.5}$$

Bearing in mind that the values of fundamental units can be chosen at random, it follows that the values of physical quantities $A, B, C, \ldots,$ considered as fixed, can vary, also at random, while values of ratios $\Pi_m, \Pi_n, \ldots$ stay unchanged. Relation (8.5) should hold true for all random values of quantities $A, B, C, \ldots$; hence, it should be independent from these parameters and therefore could be written in the simpler form (8.1).

In his book, entitled *Théorie de l'Electricité*, published in 1896, Vaschy expressed this theorem in a more general form: *any homogeneous relation between m^* physical quantities A, B, $C, \ldots, M$, N, $P, \ldots$ whose values depend on the fundamental units adopted, can be reduced to a relation between $m^* - k^*$ parameters, which are monomial combinations of A, B, $C, \ldots$ with zero dimensions* ($L^0M^0T^0 = 1$), *provided that out of the d^* fundamental units of physical quantities A, B, $C, \ldots,$ a number of k^* units of measurement could be chosen at random. It stands to reason that k^* cannot be higher than the number of fundamental units d^*.*

Vaschy did not use the symbol Π to describe his theorem. This Greek capital letter Π suggests the concept of a "product of powers", since the newly introduced quantities $\Pi_m, \Pi_n, \ldots,$ are products of powers of the original ones; they are also called "reduced quantities".

The Π theorem has been more accurately stated in 1914 [8.3] by Buckingham [1.26] and this is why it is often called Buckingham's theorem. He states this theorem as follows: *when a relation is dimensionally homogeneous, it can be reduced to a relation between a complete series of non-dimensional products* [1.26]. He describes these — non-dimensional products — monomials as Π numbers; hence the designation given by Bridgman to the principle stated by Buckingham, i.e. the Π theorem.

According to Buckingham, if m^* variables are connected through a dimensionally unknown homogeneous relation, this relation can be put in the form of a relationship between $m^* - k^*$ non-dimensional products, where $m^* - k^*$ is the number of products included in a complete series of non-dimensional products of variables. More often

[8.3] Buckingham, E., *Phys. Rev.*, IV, 4, 1914, p. 345.

than not, the number k^* is the same as the number of fundamental physical quantities of the unit system adopted. As a consequence, Buckingham — and other authors as well — considered that the non-dimensional products i^* of a complete series are equal in number to the variables m^*, less the number of fundamental physical quantities d^*, that is

$$i^* = m^* - d^*. \tag{8.6}$$

The above technical rule is convenient but not infallible; in fact, Bridgman stressed this fact in 1922 [8.4] [1.26]. One rule which can be rigorously demonstrated was stated by Van Drist in 1946 [8.5]; according to this rule: *the number of non-dimensional products within a complete series is equal to the total number of variables, less the maximal number* m_1^* *of variables which do not yield a non-dimensional product*, i.e.

$$i^* = m^* - m_1^*. \tag{8.7}$$

An equivalent, more practical rule is the following: *the number of non-dimensional products of a complete series is equal to the total number of variables* m^*, *less the rank* r^* *of the dimensional matrix*, i.e.

$$i^* = m^* - r^*. \tag{8.8}$$

This rule will be demonstrated in Section 9 of this chapter.

The Π theorem was demonstrated by Federman in 1911. In 1912, he published a report [8.6] where it was shown that this theorem was obtained as a particular case of a more general one [1.22].

More accurate formulations and demonstrations of the Π theorem were suggested, at about the same time, by Riabouchinsky [1.26] and Ehrenfest-Afanassieva [1.22]. After a comparatively short period of time, this problem was again considered by Bridgman (1933) and Esnault-Pelterie (1935—1952), who, though not changing the statement of the theorem, suggested new demonstrations, since

---

[8.4] Bridgman, P. W., *Analyse Dimensionnelle*, Yale University Press, 1932.

[8.5] Van Driest, E. R., *Journal of Applied Mechanics*, XIII, 1, March, 1946, pp. 4—34.

[8.6] Federman, A., *O nekotorîh obşcih metodah integrirovaniia uravnenii s ciastnîmi proizvodnîmi pervevo poreadka*, p. X and XI; Izvestia S. Pb. POLITEHN. inst. XVI, nr. 1, 1911 (1912).

those already available were not, in their opinion, satisfactory. New demonstrations of the Π theorem are also to be found in the recent literature. The demonstration given in [2.5] is more accurate; it is similar to that given in [1.32].

Let us consider, as suggested by the above authors, any physical phenomenon, together with the physical quantities A, B, $C \ldots$, M, N, $P, \ldots$, needed to describe the phenomenon. For simplicity it is assumed that the unknown variable is a function of only four independent variables, i.e.

$$P = f(A, B, C, D). \tag{8.9}$$

Relation (8.9) should stay invariant with the change in the units of measurement.

For the sake of the demonstration, it is assumed that the quantities, say A and B, are at the same time the fundamental physical quantities of a coherent system of units. The dimensional equations for the secondary physical quantities (B, C), will result from the following dimensional matrix:

$$\begin{array}{c|ccc|} & C & D & P \\ \hline A & a_c & a_d & a_p \\ B & b_c & b_d & b_p \end{array} \tag{8.10}$$

For all these secondary quantities we derive, from the dimensional equations (8.10), new, non-dimensional quantities:

$$\Pi_c = \frac{C}{A^{a_c} B^{b_c}}, \quad \Pi_d = \frac{D}{A^{c_d} B^{b_d}}, \quad \Pi_p = \frac{P}{A^{a_p} B^{b_p}}. \tag{8.11}$$

Replacing in (8.9) the secondary quantities by the new ones, as given from relations (8.11), results in

$$\Pi_p = f(A, B, \Pi_c, \Pi_d). \tag{8.12}$$

Relation (8.12) should also be invariant with the change in the units of measurement. Supposing such a change takes place, the value of quantities A and B become A' and B' respectively; at the same time, the values of the new quantities Π_c, Π_d and Π_p will be kept the same, as a consequence of definitions (8.11). Hence, relation (8.12) will take the form

$$\Pi_p = f(A', B', \Pi_c, \Pi_d). \tag{8.13}$$

Since the units of measurement change at random, A' and B' can take any value, which could be even constant numbers, say A_0 and B_0; hence relation (8.13) becomes

$$\Pi_p = f(A_0, B_0, \Pi_c, \Pi_d). \tag{8.14}$$

Since it is useless to find the constants explicitly, quantities A and B will be eliminated and the final relation will have the form

$$\Pi_p = \varphi(\Pi_c, \Pi_d) \tag{8.15}$$

which is the explicit form of relation (8.1).

Langhaar considers that what the Π theorem expresses is not evident. Furthermore, he insists that even Buckingham failed to come up with a rigorous proof in connection with the Π theorem and states that all that Buckingham has done is merely to present those aspects likely to make his demonstration plausible [1.26] [8.7]. To make his point, Langhaar first demonstrates some lemmas which lead him to the Π theorem. He specifically states that the logics of the Π theorem is perfectly valid only if the independent variables intervening in the problems of dimensional analysis always take positive values. When this is not the case, the non-dimensional products with fractional exponents would often be imaginary [1.26].

Comolet too thinks that the Π theorem is not mathematical evidence; this also follows from the many, more or less rigorous demonstrations already known [1.8].

In contrast with Langhaar and Comolet, Charcosset considers that the Π theorem represents a unquestionable physical truth [1.26] and consequently tried to develop a new, solid proof. His demonstra-

---

[8.7] p. 25.

tion follows a path similar to that he already suggested for the fundamental theorem of dimensional analysis already described in this work (Section I. 5.2). To arrive at the Π theorem Charcosset makes use of relation

$$f(\lambda l,\ \mu m,\ \tau t) = f(l,\ m,\ t) f(\lambda,\ \mu,\ \tau) \tag{5.27}$$

as in the demonstration which led to the fundamental theorem of dimensional analysis; he considers the two possible cases, differing from one another by the nature of the units of measurement (hence of the corresponding physical quantities) which can be :

— either fundamental; or

— fundamental and derived.

Charcosset demonstrates that the first case leads to the fundamental theorem of dimensional analysis while the second, to the Π theorem [1.26] [8.8].

In 1967, Vasilescu, a Romanian researcher, suggested another demonstration for the Π theorem [1.41] [8.9].

Since this review of Π theorem and of the various demonstrations suggested so far is, in the author's opinion, comprehensive enough for the purposes of this work, we shall end our discussion about the subject here; the reader in need of further details is referred to the works already mentioned, and in particular to [1.28].

In conclusion, we consider that the most adequate statements suggested for the Π theorem are the following :

— *any dimensionally homogeneous relation between physical quantities can be reduced to a relation involving fewer non-dimensional products;*

— *the number of non-dimensional products is equal to the number of physical quantities, less the rank of the dimensional matrix.*

8.2 **Rayleigh's Method**

Rayleigh was the first to use a new approach for the determination of non-dimensional products of powers (1899). He considered

---

[8.8] p. 205.

[8.2] Vasilescu, A. A., *A demonstration of the* Π *problem,* A report presented at the Session of the Polytechnical Institute, Galatzi, 1967.

the physical quantities $A, B, C, \ldots, M, N, P, \ldots$ arising in the development of the phenomenon under study. The dependent variable P is expressed by the function

$$P = f(A, B, C, \ldots, M, N, \ldots). \tag{3.3}$$

According to Rayleigh, function f is developed into an infinite series with respect to the powers of the independent variables, as follows:

$$P = \Sigma k_i A^{a_i} B^{b_i} C^{c_i} \ldots M^{m_i} N^{n_i} \ldots, \tag{8.16}$$

where k_i stands for the numerical factors. Of course, all terms of the infinite series (8.16) should have the same dimensional equation, namely that of the physical quantity described, P; as a consequence, one single term needs to be considered and therefore we can write

$$P = kA^a B^b B^c \ldots M^m N^n. \tag{8.17}$$

It follows that according to Rayleigh's method, one single *product of powers* is taken into consideration.

It also follows that Rayleigh's method could lead to the relation expressing the law of the evolution in time of a phenomenon. However, it should be stressed that the existence of a dimensional relation between the variables describing the physical phenomenon under consideration could not be taken as a proof for the existence of a physical law governing this phenomenon. The correlation with the physical reality can be only experimentally assessed; if however, there is a physical law, it should have the form shown in (8.17), and this is a *sine qua non* condition.

Rayleigh's method differs only formally from the procedure used in the Π theorem. The substance is the same for the two procedures, as are the computations: they lead to the same results. Nevertheless, it should be perhaps worth mentioning that the Π theorem does not make use of infinite series. While it can be considered that in a problem of physics any dependent variable could be equated to the product of powers of the independent variables multiplied by a numerical factor [1.26], it is obvious that, in practice, such an assumption should be first experimentally checked.

9 Product of Powers

9.1 Dimensional Homogeneity of a Product of Powers

From what was said above, it follows that the relationships having the general form shown in (8.17) play an important part in dimensional analysis. The product of powers of the second member of relation (8.17) also appears in ratios (8.2) which, in turn, intervene in the Π theorem, as well as in the relation written by Rayleigh between the physical quantities which describe a certain, well-defined phenomenon. Now, let us assume that the dimensional equations of variables $A, B, C, \ldots, M, N, P, \ldots$ are defined by a dimensional matrix; for simplicity, we shall use the system of units having the length L, the mass M and the time T as fundamental physical quantities:

$$\begin{array}{c|cccc|} & A & B & C\ldots & P\ldots \\ \hline \mathrm{L} & l_a & l_b & l_c & l_p \\ \mathrm{M} & m_a & m_b & m_c & m_p \\ \mathrm{T} & t_a & t_b & t_c & t_p \end{array} \tag{9.1}$$

Exponents $a, b, c, \ldots$ in the product of powers (8.17) satisfy the following conditions: *the product of powers (8.17) is dimensionally homogeneous if, and only if, the said exponents are a solution to the system of non-homogeneous, linear equations*:

$$\begin{aligned} l_a \cdot a + l_b \cdot b + l_c \cdot c + \ldots + l_m \cdot m + l_n \cdot n + \ldots &= l_p \\ m_a \cdot a + m_b \cdot b + m_c \cdot c + \ldots + m_m \cdot m + m_n \cdot n + \ldots &= m_p \\ t_a \cdot a + t_b \cdot b + t_c \cdot c + \ldots + t_m \cdot m + t_n \cdot n + \ldots &= t_p. \end{aligned} \tag{9.2}$$

In the equations above, the coefficients are the lines of the dimensional matrix (9.1); as already known, these coefficients represent dimensional exponents. To demonstrate that this condition is necessary, let us assume that P is an homogeneous dimensional product and that it has the dimensional exponents l_p, m_p and t_p; relation

(8.17) will therefore satisfy the condition of dimensional homogeneity as expressed by the known relation :

$$K_P \cdot f(A, B, C, \ldots, M, N, \ldots) = f(K_A \cdot A, K_B \cdot B, K_C \cdot C, \ldots, K_M \cdot M, K_N \cdot N, \ldots) \quad (7.15)$$

written with the new symbols. In this particular case, relation (7.15) takes the form :

$$K_P \cdot A^a B^b C^c \ldots M^m N^n \ldots = K_A^a \cdot A^a K_B^b \cdot B^b K_C^c \cdot C^c \ldots K_M^m \cdot M^m K_N^n \cdot N^n \ldots \quad (9.3)$$

From (9.3) it follows directly that :

$$K_P = K_A^a \, K_B^b \, C_C^c \ldots K_M^m \, N_N^n \ldots , \quad (9.4)$$

and from (9.4) and from (7.13) written with the new symbols $K_A = \lambda^{l_a} \mu^{m_a} \tau^{t_a}$ etc. we can easily obta in equations (9.2). Using a reversed reasoning one can easily demonstrate that equation (9.2) also represents a sufficient condition [1.26].

Of course, when P has no dimensional equation, linear homogeneous (9.2) equations will be obtained.

9.2 Independence of Non-dimensional Products of Powers

Any product of powers of the general form

$$A^a \, B^b \, C^c \ldots M^m \, N^n \ \ldots = 1 \quad (9.5)$$

and derived from (8.17) when P has no dimensional equation, is called a non-dimensional product of powers or, in short, a non-dimensional product. It is an "invariant" with respect to the fundamental units and of course, an invariant with respect to the corresponding fundamental physical quantities. The non-dimensional product of powers represented by the ratio of two variables of like nature — and hence having the same dimensional equation — is called a factor of form. This designation is due to the fact that the various factors of form contain, more often than not, geometrical quantities. However, there is no fundamental distinction between the two kinds of non-dimensional products of powers; from a dimensional point of view, they fulfil the same function. Non-dimensional products of powers are obtained by applying the Π theorem in combination with Rayleigh's method.

Let us consider the following non-dimensional products of powers:

$$\begin{aligned} \Pi_m &= A^{a_m} B^{b_m} C^{c_m} \dots \\ \Pi_n &= A^{a_n} B^{b_n} C^{c_n} \dots \\ \Pi_p &= A^{a_p} B^{b_p} C^{c_p} \dots \\ &\dots\dots\dots\dots \end{aligned} \tag{9.6}$$

The quantity exponents form the matrix below:

$$\begin{array}{c|ccc|} & A & B & C\dots \\ \hline \Pi_m & a_m & b_m & c_m \dots \\ \Pi_n & a_n & b_n & c_n \dots \\ \Pi_p & a_p & b_p & c_p \dots \\ \dots & \dots & \dots & \dots \end{array} \tag{9.7}$$

Now, what is important about non-dimensional products of powers is their independence; non-dimensional products of powers Π_m, Π_n, Π_p, ..., are called independent when no constant values k_m, k_n, k_p, ..., other than $k_m = k_n = k_p = \dots = 0$ exist, so that

$$\Pi_m^{k_m} \Pi_n^{k_n} \Pi_p^{k_p} \dots \equiv 1. \tag{9.8}$$

The necessary and sufficient condition for the Π_m, Π_n, Π_p,... *products to be independent is that the lines of the exponent matrix (9.7) be linearly independent.* This is proved by reduction to a contradiction [1.26]. The need for such a condition is shown by assuming that non-dimensional products (9.6) are independent while the lines of the exponent matrix (9.7) are linearly dependent; as a consequence, constants k_m, k_n, $k_p \ldots$ will differ from zero, so that

$$k_m \cdot i_m + k_n \cdot i_n + k_p \cdot i_p + \ldots = 0 \tag{9.9}$$

where $i = a, b. c, \ldots$ Let us now consider the self-evident relation:

$$\Pi_n^{k_m} \Pi_m^{k_n} \Pi_p^{k_p} =$$

$$= A^{k_m \cdot a_m + k_n \cdot a_n + k_p \cdot a_p + \cdots} B^{k_m \cdot b_m + k_n \cdot b_n + k_p \cdot b_p + \cdots} C^{k_m \cdot c_m + k_n \cdot c_n + k_p \cdot c_p + \cdots} \ldots \tag{9.10}$$

From (9.9) and (9.10), it follows immediately:

$$\Pi_m^{k_m} \Pi_n^{k_n} \Pi_p^{k_p} \ldots = A^0 B^0 C^0 \tag{9.11}$$

and hence

$$\Pi_m^{k_m} \Pi_n^{k_n} \Pi_p^{k_p} \ldots = 1. \tag{9.12}$$

This result contradicts the original assumption that the non-dimensional products are linearly independent.

To prove that the above mentioned condition alone is sufficient, let us assume that the non-dimensional products of powers (9.6) are not independent while the lines of the exponent matrix are linearly independent; consequently, there will be $k_m, k_n, k_p, \ldots$ constants different from zero, so that

$$\Pi_m^{k_m} \Pi_n^{k_n} \Pi_p^{k_p} \ldots \equiv 1. \tag{9.8}$$

A similar calculus will lead to:

$$A^{k_m \cdot a_m + k_n \cdot a_n + k_p \cdot a_p + \cdots +} B^{k_m \cdot b_m + k_n \cdot b_n + k_p \cdot b_p + \cdots} C^{k_m \cdot c_m + k_n \cdot b_n + k_p \cdot c_p + \cdots} \ldots = 1. \tag{9.13}$$

Since relation (9.13) is an identity with respect to variables A, B, C, ..., all exponents have a zero value; this result also contradicts the original assumption, namely that the lines of the exponent matrix are linearly independent. This proof by reduction to a contradiction confirms the validity of our statement above.

9.3 Condition of a Non-dimensional Product of Powers

Let us consider again the product of powers of the general form

$$A^a\, B^b\, C^c \ldots M^m\, N^n \ldots = 1, \tag{9.5}$$

the dimensional equations of variables $A, B, C, \ldots$ are defined by the same dimensional matrix (9.1). *The product of powers (9.5) is non-dimensional if, and only if, the quantity exponents are a solution of the system of linear, homogeneous equations:*

$$\begin{aligned} l_a \cdot a + l_b \cdot b + l_c \cdot c + \ldots + l_m \cdot m + l_n \cdot n + \ldots &= 0 \\ m_a \cdot a + m_b \cdot b + m_c \cdot c + \ldots + m_m \cdot m + m_n \cdot n + \ldots &= 0 \\ t_a \cdot a + t_b \cdot b + t_c \cdot c + \ldots + t_m \cdot m + t_n \cdot n + \ldots &= 0 \end{aligned} \tag{9.14}$$

The coefficients in equations (9.14) are the lines of the dimensional matrix (9.1); this condition follows immediately from that stated before (Section II. 9.1).

As known [2.8], any system of homogeneous equations admits a fundamental system of solutions. In the case of equations (9.14), the fundamental system of solutions leads to $m^* - r^*$ linearly independent series of values for the quantity exponents; symbol r^* represents the rank of the dimensional matrix corresponding to the system of equations (9.14). Since the quantity exponents are linearly independent, based on the condition imposed before (Section II. 9.2), they will lead to independent non-dimensional products of powers. A number of independent non-dimensional products of powers higher than $m^* - r^*$ cannot exist, since equations (9.14) do not admit more than $m^* - r^*$ linearly independent solutions. Hence the following statement: *any fundamental system of solutions for equations (9.14) will supply the quantity exponents for a complete series of non-dimensional products of powers corresponding to variables $A, B, C, \ldots$ $\ldots, M, N, \ldots$* The reverse is also true: *the quantity exponents of a complete series of non-dimensional products corresponding to variables $A, B, C, \ldots, M, N, \ldots$ are a fundamental solution of the system of equations (9.14).*

9.4 The Number of Non-dimensional Products of Powers

As a corollary to the statement above, we can make the following one : *the number of products of a complete series of non-dimensional products of powers corresponding to m^* variables $A, B, C, \ldots, M, N, \ldots$ is $m^* - r^*$*, where r^* is the rank of the dimensional matrix of these variables.

According to the Π theorem, the choice of quantities considered as fundamental out of the quantities of a given (3.2) function is made at random. As a consequence, various series of $m^* - r^*$ non-dimensional products of powers might result. A complete series made up with $m^* - r^*$ products may yield other complete series of powerless products, different from the original ones. The conversion of non-dimensional products of powers is operated by means of elemental algebraic operations. It leads to the replacement of function $\varphi(\Pi_1, \Pi_2, \Pi_3 \ldots) = 0$ by function $\varphi(Q, R, S, \ldots) = 0$, where symbols $Q, R, S, \ldots$ represent new, conveniently built up non-dimensional products of powers, having, say, the form $Q = \Pi_1, \Pi_2$; $R = \Pi_1 \Pi_3^2$; $S = \Pi_2^2 \Pi_3^{-1}, \ldots$ Conversion of non-dimensional products of powers might be determined either by the need for eliminating a quantity or by the desire to operate with classic non-dimensional products of powers — Reynolds, Nusselt etc. — and, generally, by the convenience of experimental measurements. Any conversion of the like can be made only if the number of non-dimensional products of powers is the same as that of the old ones. This replacement is justified by the fact that relation $\rho(\Pi_1, \Pi_2, \Pi_3, \ldots)=0$ is unknown and has to exist. Any relation between $\Pi_1, \Pi_2, \Pi_3 \ldots$ implies a relation between $Q, R, S, \ldots$, and vice-versa. Each series of new or old non-dimensional products of powers should be complete.

9.5 Condition of a Dimensional Product of Powers

Any product of powers of the general form

$$P = A^a B^b C^c \ldots M^m N^n \ldots \tag{8.17}$$

where P has a dimensional equation and is called a dimensional product of powers or, in short, product of powers. Such products intervene whenever Rayleigh's method is applied.

Let us consider a product of powers of the form (8.17); in the dimensional matrix (9.1), the constant values of quantity exponents are not all equal to zero. The following question arises [1.26] : which is the condition of the existence of a relation of the form (8.17)? The answer shall be put as follows : *if P has a dimensional equation, a dimensionally homogeneous product of the form (8.17) will exist if, and only if, the dimensional matrix of variables A, B, C, ..., M, N, ... is of the same rank as the dimensional matrix of variables A, B, C, ..., M, N, P, ...* This condition expresses the idea that equations (9.2) are compatible since quantity exponents *a, b, c, ...* ... *m, n...* (if any) are a solution to equation (9.2). In algebra, it is demonstrated that equations (9.2) are compatible if, and only if, the rank of the coefficients l_a, m_a, t_a, ..., t_n, ... matrix does not change when column l_a, m_p, t_p is added to the matrix.

9.6 Theorem of the Restricted Dimensional Analysis

From what has been said above, we can prove that *if the function*

$$P = f(A, B, C, \ldots, M, N, \ldots) \tag{3.3}$$

is dimensionally homogeneous and, if P has a dimensional equation, then there will be a product of powers built up with variables A, B, C,... ... M, N,... and having the same dimensional equation as P. For demonstration purposes, let us assume that relation (3.3) is dimensionally homogeneous and that there are no products of powers of variables *A, B, C, ..., M, N, ...* with the same dimensional equation as *P*; let us further assume that the rank of dimensional matrix (9.1) is r^*. From what was said in Section II.9.5 and assuming the non-existence of products of powers of variables *A, B, C, ...* ..., *M, N, ...* having the same dimensional equation as *P*, we can state that the rank of the matrix obtained by suppressing the last column of the dimensional matrix (9.1) is lower than r^*; as a consequence, in the r^* right-hand columns of the dimensional matrix there will be one r^* order determinant different from zero [1.26].

Let us consider the particular case : $r^* = 3$. From what was said above, we may write

$$\Delta = \begin{vmatrix} l_a & l_b & l_p \\ m_a & m_b & m_p \\ t_a & t_b & t_p \end{vmatrix} \neq 0 \cdot \tag{9.15}$$

Noting by α, β, γ the corresponding minors of determinant (9.15) development with respect to the last column, we can write

$$\Delta = l_p \cdot \alpha + m_p \cdot \beta + t_p \cdot \gamma \neq 0. \tag{9.16}$$

On the other hand, the determinant

$$\Delta' = \begin{vmatrix} l_a & l_b & l_i \\ m_a & m_b & m_i \\ t_a & t_b & t_i \end{vmatrix} \tag{9.71}$$

where $i = a, b, c, \ldots$, also developed with respect to the last column leads to

$$\Delta' = l_i \cdot \alpha + m_i \cdot \beta + t_i \cdot \gamma. \tag{9.18}$$

For $i = a$ and $i = b$, the determinant (8.18) takes a zero value since it has two identical columns. For $i = c, d, \ldots$, the determinant will again take a zero value since, as already shown, the rank of the dimensional matrix of variables $A, B, C, \ldots, M, N, \ldots$ is lower than 3. Hence

$$l_i \cdot \alpha + m_i \cdot \beta + t_i \cdot \gamma = 0 \tag{9.19}$$

and this will always be true.

According to the assumption accepted about the dimensional homogeneity of relation (3.3), relation (7.15) will be an identity with respect to variables λ, μ, τ; as a consequence :

$$\lambda = H^\alpha, \quad \mu = H^\beta, \quad \tau = H^\gamma \tag{9.20}$$

where H is an arbitrary positive constant. Bearing in mind relations (7.13), relation (9.19) leads to

$$K_i = 1 \tag{9.21}$$

where $i = a, b, c, \ldots$ As a consequence, relation (7.15) takes the form

$$K_p \cdot P = f(A, B, C, \ldots, M, N, \ldots) \tag{9.22}$$

where

$$K_p = H^{t_p \cdot \alpha + m_p \cdot \beta + l_p \cdot \gamma} \tag{9.23}$$

is a constant; this constant is arbitrary since H is also an arbitrary constant and according to (9.16) the exponent differs from zero. Relation (9.22) shows therefore that there is no connection between variables $A, B, C, \ldots, M, N, \ldots$ and variable P, i.e. $f(A, B, C, \ldots$ $\ldots, M, N, \ldots)$ is not a function. But this conclusion contradicts our original assumption according to which $P = f(A, B, C, \ldots$ $\ldots, M, N, \ldots)$; and hence we have proved the existence of a product of powers built up with variables $A, B, C, \ldots, M, N$, and having the same dimensional equation as P.

From knowns, the above demonstration is made for the particular case when $r^* = 3$. If the rank of dimensional matrix r^* is lower than 3, then only those lines of the dimensional matrix holding a r^* order determinant different from zero will be taken into consideration. In such a case, the proof is made as before, except that terms λ, μ, τ, corresponding to the lines suppressed from the dimensional matrix, are equal to unity.

10 Restricted Dimensional Analysis

10.1 Short History

Problems of dimensional analysis were always a concern of scientists. A critical review of dimensional analysis through the centuries is given in [1.23]. Euclid, Ptolemy, Pappus, Gallilei, Descartes, Newton, Leibniz, Euler, Lagrange, Laplace, Stokes, Reynolds, Planck, Einstein have all dealt with dimensional analysis. The principles of dimensional analysis were stated in the eighteenth century. Its bases were laid for the first time by Fourier, în 1822. Indeed, Fourier drew attention to the fact that in the physical relations expressed in the Gaussian system of units, all terms of the same sum have the same dimensions while the arguments of transcendent (trigonometric, logarithmic) functions are non-dimensional; this property constitutes the dimensional homogeneity of physical relationships. Fourier applied the geometric concept of *dimension* to physical quantities. In 1848 [10.1], Bertrand combined the physical quantities and obtained non-dimensional products of powers. In 1871, Maxwell introduced a novelty into dimensional analysis, namely the concept of *dimensional equation*. Bertrand, in 1878 [10.2] added a specific condition to the statement made by Fourier which now runs as follows: *the condition of dimensional homogeneity required for the physical relationships should be satisfied by the still unknown formulae as well; if the physical quantities that the latter include are known, such relations can often be derived on the basis of this condition*. In 1890, Vaschy stated the Π theorem, which is of a particular importance in dimensional analysis. Rayleigh applied for the first time, in 1899, the non-dimensional product of powers in fluid mechanics and expressed a dependent physical quantity by a product of powers of the independent physical quantities involved in the evolution of a certain physical phenomenon. In 1922, Bridgman made important contributions with regard to the number of non-dimensional products of powers.

In view of their respective contributions, Palacios suggests that Fourier should be considered as the forerunner of dimensional analysis, and Rayleigh, its founder; Palacios also states that Bridgman is the greatest authority in the field [1.32] [10.3].

In his book published in 1952 [10.4], Huntley described his *method of components of fundamental dimensions* which refers to the orientation of the fundamental physical quantity of length in accordance with a cartesian system of axes; according to Palacios, this method is the most important progress made in this field since Fourier and is extremely useful in practical applications [1.32] [10.5].

[10.1] Bertrand, J., Note sur la similitude en mécanique, *Journ. de l'Ecole Polytechnique*, Cahier 32, 1848.

[10.2] Bertrand, J., Sur l'homogénéité dans les formules de physique, *Comptes Rendus de l'Académie des Sciences*, LXXXVI, 15 avril 1878.

[10.3] pp. VII, VIII, XV.

[10.4] Huntley, H. E., *Dimensional Analysis*, Macdonald, London, 1952.

[10.5] p. 59.

In his book published in 1951[10.6] Langhaar gave the proof of the theorem of the restricted dimensional analysis; according to this theorem, any monomial, dimensionally homogeneous physical relationship has the form of a product of powers; this form was used, long before, by Rayleigh also. The 1956 French edition of Langhaar's work [1.26] includes an appendix where Charcosset points out that, since the older demonstration fails to lead to the correct statement of the fundamental theorem of dimensional analysis, he will suggest a new demonstration.

It is perhaps worthwhile mentioning that important as it is one cannot say when and who stated the fundamental theorem of dimensional analysis for the first time.

Using dimensional analysis, Einstein obtained an important result in physics; indeed, in 1907, he worked out the expression for the atom oscillating frequency within a solid. Other applications of dimensional analysis were tried by Vaschy, Rayleigh, Riabouchinsky, Bridgman. In technics, it was correlated with similarity, which opens a wide field of applications to dimensional analysis.

Problems of dimensional analysis were published in several works, of which we mention those written by: Bridgman (1922—1949), Esnault-Pelterie (1946 — 1952), Martinot-Lagarde (1946), Langhaar (1961), Duncan (1953), Palacios (1956), Comolet (1958), Huntley (1967), Douglas (1969).

10.2 Definition of Dimensional Analysis

The dictionaries and/or lexicons offer no definition of dimensional analysis. In the most recent works published on this subject, for instance in [2.29], the term *analysis* is usually described as: the search for the component parts of a given object, event or phenomenon; or else the breaking down of a complex into its component elements. While no description is offered for the term *dimensional* in [2.29] the term *dimension* — from which *dimensional* comes — is described as: a number expressing the connection between a derived unit and the fundamental units from which it is derived. The term dimensional is used, however, in the expression *dimensional equation*, meaning a relation expressing a physical quantity as a function of the

[10.6] Langhaar, H. L., *Dimensional Analysis and Theory of Models*, John Wiley, New York, 1951.

fundamental quantities of a system of units; thus, for instance, the

$$[F] = \mathrm{LMT}^{-2} \tag{5.44}$$

relation represents the dimensional equation for force, as a physical quantity in the system with three fundamental units i.e.: length L, mass M, and time T. However, often dimensional equation is used, as known, for dimension; obviously, this latter concept differs from the description given above [2.29] for the term dimension.

Several suggestions for a definition of dimensional analysis could be found in the works published in this field. Thus, Saint-Guilhem states [1.35] [10.7] that the object of dimensional analysis is to investigate the form of an equation taking as a criterion the homogeneity conditions. Kirpicev [1.22][10.8] states that the dimensional analysis is a method applied to establish, on a dimensional basis, the similarity criteria contained by the criterial equation; he also states [1.22][10.9] that the dimensional analysis represents: the unique method available to the establishment of similarity criteria when no relating equation can be written, and here lies its great importance. He also points out that the designation of *similarity criterion* is synonymous to that of *non-dimensional product of powers*. In his book [1.26][10.10] Langhaar states that dimensional analysis is the study of the general form of equations describing natural phenomena, its object being to obtain information on a physical phenomenon on the basis of the simple assumption that the said phenomenon can be described by means of a dimensional equation which should be true when referred to certain variables; he also maintains that the result of a dimensional analysis is to reduce the number of variables involved in the problem dealt with [1.26][10.11]. Langhaar further states that dimensional analysis has become an important mathematical tool in the hands of the experimental researcher [1.26] [10.12]. According to Comolet [1.8] [10.13] dimensional analysis is the study of the general form of equations describing natural phenomena. Palacios [1.32] does not agree with Bridgman's assertion, namely that dimensional analysis serves to establish the necessary limits of a possible relationship between the variables of

---

[10.7] Chapter 6.
[10.8] p. 1331.
[10.9] p. 1336.
[10.10] p. XV.
[10.11] p. 1.
[10.12] p. 2.
[10.13] p. 1.

a physical system even though accurate and detailed information about the fundamental equations on which the solution should rely were not available[10.14]. As a consequence, Palacios says that if Bridgman's assertions were true, then dimensional analysis would permit us to make correct predictions on phenomena whose fundamental laws are not known [1.32][10.15]. Jacovleff [2.14] [10.16] states that the thermodynamic similarity concept pertains to that mathematical field known in Europe under the designation of theory of similarity and in America under the designation of dimensional analysis. Speaking about the same problem, Adams [2.1][10.17] states that, admitting that all factors which describe a certain physical state were known, the dimensional analysis is the method whereby information gathered on the said factors can be ordered so that it could be used to plan the future experiments and to interpret the relevant results; he further says that this method is particularly suitable when mathematical relationships are not known, or when they are too complex, and concludes that dimensional analysis will yield in this case a logic group of factors under the form of non--dimensional combinations. Pawlowsky [2.20][10.18] maintains that the object of dimensional analysis is to supply information on relationships existing between measurable physical quantities used to describe physical phenomena, and further: these relationships are sometimes a sufficient basis to predict the law governing a certain phenomenon; he also considers that dimensional analysis aims at grouping the physical quantities in non-dimensional products so that experimental results could be expressed through relationships between these products. According to Brun and Martinot-Lagarde [2.5][10.19] dimensional analysis is called the method of similarities investigation based on Vaschy's theorem.

In the author's view, *dimensional analysis* is the study of relationships which describe physical phenomena. On the basis of the definition for the term *analysis*, of its correlation with the term *dimensional equation* and of the accepted synonyms *dimensional analysis or dimension*, the author prefers to state that: dimensional analysis is a method whereby monomial, dimensionally homogeneous

---

10.14 Bridgman, P. W., *Encyclopaedia Britanica*: Dimensional Analysis, 1951.
10.15 p. XV.
10.16 Chapter 9, point 2.
10.17 p. 140.
10.18 p. 187.
10.19 p. 154.

relationships are established between measurable variables which describe physical phenomena. Dimensional analysis can be applied to the study of a variety of problems raised by physical quantities, units of measurement, fundamental quantities, physical relationships and/or the numerical factor, namely :

— conversion of the value of a physical quantity or numerical factor when the units of measurement are changed ;

— assessment of the physical significance of an element of a physical relationship, i.e. quantity or term ;

— deriving the conversion relations for the units of measurement or the fundamental quantities ;

— checking the dimensional homogeneity of physical relations ;

— assessment of non-dimensional products of powers, when no equation describing physical phenomena, but only the measurable variables intervening in their description, are available.

In the literature, the expression "conventional dimensional analysis" is sometimes used, as for instance in [1.30] [10.20]. The special issue of the *Journal of the Franklin Institute*, December 1971, dedicated to "dimensional analysis and similarity", is entitled "Modern Dimensional Analysis", the editors thereby stressing the difference between old and recent information and concepts available on the subject. It is the author's opinion that the new data supplied on this subject refer mostly to the formal aspects of dimensional analysis, i.e. new, more accurate mathematical demonstrations and new theorems on the conventional dimensional analysis.

10.3 Application of the Π Theorem and of Rayleigh's Method

Let us consider the function of physical quantities

$$f(A,\ B,\ C,\ \ldots,\ M,\ N,\ P,\ \ldots) = 0 \qquad (3.2)$$

arising in the progress of a given physical phenomenon. According to Π theorem, relation (3.2) may be written :

$$\varphi(\Pi_m,\ \Pi_n,\ \ldots) = 0, \qquad (8.1)$$

[10.20] p. 424.

where $\Pi_m, \ldots$ represents $m^* - r^*$ non-dimensional products of powers. On the other hand, in compliance with Rayleigh's method, relation (3.2) is written in the form

$$P = k\, A^a\, B^b\, C^c \ldots M^m\, N^n \ldots \tag{8.17}$$

If we choose to use Π theorem for the assessment of non-dimensional products of powers, several methods are known. One of them was applied for the first time by Bertrand in an attempt to combine certain quantities so as to obtain non-dimensional products of powers. Appel's methods pays due consideration to the known fact that the change in the units of measurement should not affect physical relationships [10.21]. Federman devised a method based on his own demonstration of the Π theorem. Tolman also considers the change in the units of measurement; to this end, he considers the particular case in which the unit conversion factors are equal [10.22]. The method devised by Buckingham is also based on the demonstration offered by himself for the Π theorem. Obviously, since the quantities selected as fundamental among those contained in relation (3.2) represent an arbitrary choice, a variety of series of $m^* - r^*$ non-dimensional products of powers may result. This problem is studied in connection with the use of non-dimensional products of powers in physical similarity.

In order to apply the Π theorem to relation (3.2) let us write the following ratios:

$$\Pi_m = \frac{M}{A^{a_m}\, B^{b_m}\, C^{c_m}}, \quad \Pi_n = \frac{N}{A^{a_n}\, B^{b_n}\, C^{c_n}}, \ldots \tag{8.2}$$

Now we introduce in the above relation the dimensional equations written for the quantities involved in relation (3.2), as given by the dimensional matrix:

$$\begin{array}{c|cccc|} & A & B & C\ldots & M\ldots \\ \hline \mathrm{L} & l_a & l_b & l_c & l_m \\ \mathrm{M} & m_a & m_b & m_c & m_m \\ \mathrm{T} & t_a & t_b & t_c & t_m \end{array} \tag{10.1}$$

[10.21] Appel and Domeville, *Mécanique Théorique*, 1912.

[10.22] Tolman, R. C., Principle of similitude, *Phys. Rev*, III, 1914, VI, 1915, VII, 1916.

The condition for the dimensional homogeneity of a non-dimensional product of powers Π_m is expressed by the system of non-homogeneous linear equations:

$$
\begin{aligned}
l_a \cdot a_m + l_b \cdot b_m + l_c \cdot c_m &= l_m \\
m_a \cdot a_m + m_b \cdot b_m + m_c \cdot c_m &= m_m \\
t_a \cdot a_m + t_b \cdot b_m + t_c \cdot c_m &= t_m
\end{aligned}
\tag{10.2}
$$

which is written for each (8.2) ratio defining a non-dimensional product of powers. The system is always determined and admits a unique solution. According to theorem Π, the result of dimensional analysis computations, as applied to relation (3.2), will be a relation which may have the general form

$$\Pi = \varphi(\Pi_m, \Pi_n, \ldots). \tag{10.3}$$

If Rayleigh's method is applied, then relation (8.17) and the dimensional equations given by the dimensional matrix (10.1) should be taken into consideration and introduced in relation (8.17). Now, the condition for the dimensional homogeneity of the product of powers (8.17) is expressed by the system of non-homogeneous linear equations:

$$
\begin{aligned}
l_a \cdot a + l_b \cdot b + l_c \cdot c + \ldots + l_m \cdot m + l_n \cdot n + \ldots &= l_p \\
m_a \cdot a + m_b \cdot b + m_c \cdot c + \ldots + m_m \cdot m + m_n \cdot n + \ldots &= m_p \\
t_a \cdot a + t_b \cdot b + t_c \cdot c + \ldots + t_m \cdot m + t_n \cdot n + \ldots &= t_p.
\end{aligned}
\tag{9.2}
$$

The number e^* of distinct (9.2) equations which can be written for a given physical phenomenon may be at the most equal to the number d^* of fundamental quantities providing the system of units adopted at random. Hence, we shall always have

$$e^* \leqslant d^*. \tag{10.4}$$

The number n^* of unknowns in equations (9.2) will always be lower by one unit than the number m^* representing the quantities used to describe the physical phenomenon under consideration. Hence

$$n^* = m^* - 1 \tag{10.5}$$

and this is always true.

On the other hand, as known, the systems of units now in use contain 2 to 4 fundamental quantities; hence

$$e^* \leqslant 2\ldots4. \tag{10.6}$$

However, more often than not, 5 to 7 physical quantities are used to describe physical phenomena; consequently

$$n^* = 5\ldots7; \tag{10.7}$$

which however does not imply that their number cannot exceed these limits.

Consequently, two situations may arise for the $e^* - n^*$ correlation, namely:

— the determination situation, when $e^* \geqslant n^*$;

— the non-determination situation, when $e^* < n^*$.

In the first case, the system of equations (9.2) admits only one solution for the quantity exponents $a, b, c, \ldots, m, n, \ldots$ and relation (8.17) is univocally determined up to the value of the numerical factor of proportionality k. This situation corresponds to $m^* - r^* = 1$. In the second case, the system of equations (9.2) is solved by admitting some unknowns as independent variables; hence, an infinity of solutions for the quantity exponents $a, b, c, \ldots, m, n, \ldots$ and the same number of (8.17) relations, all of which are dimensionally homogeneous. This situation corresponds to $m^* - r^* > 1$. Here again it is obvious that, owing to the fact that the unknowns considered as independent variables are chosen at random from the quantity exponents of relation (8.17), a variety of series of $m^* - r^*$ non-dimensional products of powers may result. This problem is important if the aim is to apply the non-dimensional products of powers to physical similarity.

If Rayleigh's method is applied, the result of the dimensional analysis is a relation which may be of the form, say,

$$P = kA^{a_1}\, B^{b_1}\, C^{c_1} \ldots \tag{10.8}$$

or

$$\Pi = k\Pi_m^{k_m}\, \Pi_n^{k_n} \ldots, \tag{10.9}$$

where a new non-dimensional product of powers was introduced. Obviously, relation (10.3) derived above by applying the Π theorem may also be written in the form (10.9), but here the introduction of the numerical coefficient factor of proportionality, k, is mandatory. Therefore, the Π theorem and Rayleigh's method have both led to relations of the form (10.9).

11 Determination of the Factor and of the Exponent Values

11.1 The Graphic Method

Dimensional analysis conducted according to the Π theorem or the Rayleigh's method always leads to relationships of the general form: (10.9). The numerical factor of proportionality, k, and the exponents of non-dimensional products of powers m, n, ... have values that cannot be assessed by means of restricted dimensional analysis; they are assessed on the basis of data produced either by experimental measurements or by the theory of the phenomenon under consideration.

Let us consider the case when only one non-dimensional product of powers is obtained, that is $m^* - r^* = 1$; in such a case, relation (10.9) takes the form

$$\Pi = k, \tag{11.1}$$

the value of factor k resulting straightforwardly from the values of variables intervening in the non-dimensional product of powers Π. In such a case, the values of theoretically corresponding variables should be assessed by one single experimental measurement.

Assuming that $m^* - v^* = 2$, relation (10.9) takes the form

$$\Pi = k\Pi_1^m. \tag{11.2}$$

Applying relation (11.2) the logarithmic function results in

$$\ln\Pi = \ln k + m\ln\Pi_1 \tag{11.3}$$

and operating some obvious substitutions, we can write relation (11.3) in the form

$$y = mx + n. \tag{11.4}$$

When displayed on a system of rectangular axes, relation (11.4) will represent a straight line, m being its slope and n, the distance from the origin to the intersection of the straight line with the ordinate axis. For this graphic display, the values of non-dimensional products of powers Π and Π_1, computed on the basis of data produced by at least two experimental measurements, are needed. These values will be recorded on the diagram with $\ln \Pi$ on the ordinate and $\ln \Pi_1$ on the abscissa. On the basis of this diagram, the following relations may be written

$$m = \operatorname{tg} \alpha = m_1 = \frac{\ln \Pi'' - \ln \Pi'}{\ln \Pi_1'' - \ln \Pi_1'} \tag{11.5}$$

where α is the angle of the straight line with the abscissa axis; the value of angle α, or else the data needed for the right-hand ratio of relations (11.5), as well as the value of the constant will be read on the diagram; the value of factor k is readily computed. This value can also be estimated with the obvious relation

$$k = \frac{\Pi}{\Pi_1^{m_1}} \cdot \tag{11.6}$$

Should the experimental points follow a curved path, a broken line could be admitted; values k and m thus obtained vary with the various sections of the broken line.

When $m^* - r^* = 3$, relation (10.9) becomes

$$\Pi = k \Pi_1^m \Pi_2^n. \tag{11.7}$$

Keeping one of the non-dimensional products of powers at a constant value, say $\Pi_1 = \text{const.}$, relation (11.7) takes the form

$$\Pi = K \Pi_2^n \tag{11.8}$$

where, obviously, the new factor

$$K = f(k, \Pi_1, m). \tag{11.9}$$

Relation (11.8) is of the same form as (11.2) and, therefore, applying the logarithmic function, it will be described by a straight line. Now, we also ascribe to Π_1 other values, so as to obtain a family of straight lines ; on each straight line the non-dimensional product of powers Π_1 will have a constant value, yet different from one straight line to another. On the basis of experimental measurements, the relevant diagrams are drawn up and from these diagrams the values of unknowns intervening in relation (11.7) might be extracted. With the same end in view — i.e. to find the values for the factor and exponents of relation (11.7) — random values are chosen for one exponent, say $m = -m_1$. Relation (11.7) becomes

$$\Pi \Pi_1^{-m_1} = k \Pi_2^n ; \tag{11.10}$$

which is the same in form as (11.2) and, by applying the logarithmic function, will be represented by a straight line. Data obtained from experimental measurements are used to compute non-dimensional products of powers and values $\ln \Pi \ \Pi_1^{-m_1}$ will be recorded on the ordinate while values $\ln \Pi_2$, on the abscissa axis. Now we consider the diagram where the representative curve is a straight line and we extract the data which will lead to the values of the unknowns.

In general, it is possible that $m^* - r^* > 3$, i.e. relations of the form

$$\Pi = k \Pi_1^m \Pi_2^n \Pi_3^p \Pi_4^q \ldots \tag{11.11}$$

will be obtained. The non-dimensional products of powers are conveniently grouped and random values are ascribed to some exponents so that relation (11.11) will take either form (11.2) or form (11.7); as shown before, data resulting from experimental measurements are used to obtain the values for the unknowns. Of course, the amount of the calculation will increase with the number of non-dimensional products of powers ; when the $m^* - r^*$ difference is wide, computations can be mechanized.

11.2 The Algebraic Method

The values of unknowns k, m, n, ... from relation (10.9) — obtained by dimensional analysis — can be obtained by algebraic means too. Let us consider, for instance, the relation

$$\Pi = k \Pi_1^m \Pi_2^n \tag{11.7}$$

including three unknowns. For the variables defining the non-dimensional products of powers we can use the data obtained from at least three experimental measurements, then compute the values for Π_1, Π_2 and Π_3. On this basis, we can write the system of three equations with three unknowns:

$$\begin{aligned}\Pi' &= k\Pi_1'^m\Pi_2'^n\\ \Pi'' &= k\,\Pi_1''^m\,\Pi_2''^n\\ \Pi''' &= k\,\Pi_1'''^m\,\Pi_2'''^n.\end{aligned} \tag{11.12}$$

The solution of the system of equations (11.12) will be the values of k, m, n. The algebraic method can be applied irrespective of the number of non-dimensional products of powers; hence, it is generally valid.

Whenever the methods devised to establish the values for the numerical factor of proportionality and exponents of non-dimensional products of powers are applied in practice, one should use a higher number of experimental data than the minimum needed. Taking into account the relations expressing non-dimensional products of powers, as well as the values found for factor k and exponents m, n, ..., the relations between non-dimensional products of powers (11.7) become relations between physical quantities.

★

For some physical phenomena, the form of relation (11.7) can be determined without using experimental measurement data. To this end, all that is needed is to pay due consideration to the theory of the phenomenon under study, i.e. to its physical model. According to [1.32] [11.1], this approach can be used in many cases.

[11.1] p. 49.

III. Applications of the Restricted Dimensional Analysis

12 Introductory Remarks

12.1 Data which can be Obtained by Dimensional Analysis

Before presenting some examples of dimensional analysis applications, a few comments will be made on the actual situations which could be met in practice.

The fact that in a certain physical phenomenon m^* variables $A, B, C, \ldots, M, N, P, \ldots$ arise can never be over-emphasized. These variables can be physical quantities, universal constants or characteristic constants; they always have their own dimensional equations. The variables should be consistent with the theory of the physical phenomenon they describe, the value of one variable being determined by the values of other variables. This should be either a scientific fact or it might be admitted as a working hypothesis. From the Π theorem, the function of m^* variables used to describe the phenomenon

$$f(A, B, C, \ldots, M, N, P, \ldots) \tag{3.2}$$

may be written as a function of i^* non-dimensional products of powers:

$$\varphi(\Pi_m, \Pi_n, \ldots) = 0. \tag{8.1}$$

There is always the possibility that within a complete series of independent non-dimensional products of powers the independent variable, say P, will intervene only in one non-dimensional products of powers; as a consequence

$$P = kA^aB^bC^c \ldots (\Pi_m, \Pi_n, \ldots). \tag{12.1}$$

Dimensional analysis is that method by which new data can be obtained on physical phenomena, starting from the simple assumption that any phenomenon can be correctly described by any small number of variables interconnected through a dimensionally homogeneous relationship. This fact explains Langhaar's statement, i.e.: the general character of the method is at the same time its force and its weakness [1.26][12.1].

It follows from (3.2) and (8.1) that the result of applying dimensional analysis to a problem is a reduction in the number of variables since always $i^* < m^*$. Now, the following question will arise: what is the benefit of cutting down the number of variables?

The answer is obvious if we only care to consider the way in which a physical relationship based on experimental data is derived. Any function of one variable can be diagramatically represented by a curve. A function of two variables will be represented by a family of curves, that is an alignment chart; one curve is needed for each of the second variable values. The function of three variables will then be represented by a series of alignment charts and for each value of the third variable an alignment chart will be needed. A function of four variables will be represented by a series of alignment charts a.s.o. Now, let us assume that for each curve drawn on the diagram, five experimental points should be first determined; then, for an alignment chart including five curves, 25 experimental curves will be needed, while for a set of five similar alignment charts, 125 experimental points will be needed. Obviously, the reduction in the number of variables — a task which can be achieved using dimensional analysis — will lead to a greatly simplified diagrammatical representation of results obtained from experimental measurements; clearly, dimensional analysis is an extremely useful tool for experimental research.

[12.1] p. 1.

Relation (8.1) written between the non-dimensional products of powers can hold either one single product, $i^* = 1$ or several such products : $i^* > 1$; this feature is worth noting when it comes to estimating those data which can be obtained by dimensional analysis.

Obviously, the most complete data will be those obtained when the number of non-dimensional products of powers has the unity value ; in such a case, the relevant product will be equal to a constant, that is a pure number. Relation (8.1) will then be reduced to a relationship among physical quantities, i.e.

$$P = kA^a B^b C^c \ldots \tag{8.17}$$

When the number of non-dimensional products of powers is greater then unity, the φ function cannot be derived by dimensional analysis. As is shown, in such cases, we can have two alternatives : either to use data obtained from experimental measurements as these could lead to the values of the numerical factor and of exponents of non-dimensional products of powers (Sections II.11.1,2) ; or else to use the theory of the phenomenon, which could yield only the values of exponents of non-dimensional products of powers, while the value of the numerical factor will be obtained on an experimental basis. Anyhow, the final result will still be a relationship between physical quantities.

To illustrate what was said above, let us consider the physical phenomenon of a simple pendulum oscillating in a gravitationaly uniform field ; this example of a dimensional analysis application is the one most often given in the literature. As mentioned before (Section I. 3.1) this phenomenon involves the following variables : pendulum length l, pendulum mass m, oscillation amplitude α and time t. The only intervening force is weight G, given by the relation

$$G = mg, \tag{12.2}$$

where g is the normal acceleration of the free fall.

Based on what was said before, the implicit function of $m^* = 4$ variables will be written

$$f(l, m, \alpha, t, G) = 0. \tag{12.3}$$

Bearing in mind the system of units with three fundamental quantities commonly used in mechanics, the dimensional equations of the variable will be derived from the following dimensional matrix :

$$\begin{array}{c|cccc|} & l & m & t & G \\ \hline \mathrm{L} & 1 & 0 & 0 & 1 \\ \mathrm{M} & 0 & 1 & 0 & 1 \\ \mathrm{T} & 0 & 0 & 1 & -2 \end{array} \tag{12.4}$$

The rank of the dimensional matrix (12.4) is $r^* = 3$.

The number of non-dimensional products of powers is $i^* = m^* - r^* = 4 - 3 = 1$.

Now, we shall resort to Rayleigh's method and write the explicit relation of the independent variable

$$t = kl^a m^b G^c \rho\,(\alpha), \tag{12.5}$$

due consideration being given to the fact that α has no dimensional equation.

Dimensional equations derived for the variables from the dimensional matrix (12.4) are introduced in relation (12.5). The dimensional homogeneity condition is expressed by the system of non-homogeneous linear equations :

$$\begin{aligned} a + c &= 0 \quad (\mathrm{L}) \\ b + c &= 0 \quad (\mathrm{M}) \\ -2c &= 1 \quad (\mathrm{T}). \end{aligned} \tag{12.6}$$

The system of equations (12.6) admits the solution :

$$a = \frac{1}{2}, \quad b = \frac{1}{2}, \quad c = -\frac{1}{2}, \tag{12.7}$$

which lead to the following matrix :

$$\begin{array}{c|cccc|} & a & b & & c \\ & l & m & t & G \\ \hline \Pi & -\frac{1}{2} & -\frac{1}{2} & 1 & -\frac{1}{2} \end{array} \tag{12.8}$$

Based on matrix (12.8), the relation expressed by physical quantities will be written as

$$t = k\sqrt{\frac{lm}{G}}\,\varphi(\alpha). \tag{12.9}$$

It follows from (12.2) and (12.9) that

$$t = k\,\frac{l}{g}\,\varphi(\alpha) \tag{12.10}$$

which is correct, the complete relationship being

$$t = 2\pi\sqrt{\frac{l}{g}}\left(1 + \frac{\alpha^2}{16}\,\ldots\right)\cdot \tag{12.11}$$

For physical phenomena whose number of non-dimensional products of powers is above unity, relation (8.1) in the explicit form (12.1), is fruitful in physical similarity; that is to say if relation (12.1) is applied to two similar physical systems — i.e. two physical systems where the non-dimensional products of powers of the first physical system have the same values as their homologous non-dimensional products of powers of the second physical system — then function φ should take the same value for both physical systems. As a consequence, all that is needed is to find the value of function φ for one of the two physical systems; this can be done on the basis of values found for variables $A, B, C, \ldots M, N, P, \ldots$ in any particular case, say by using a low-scale model.

But there are other ways of establishing a relationship among variables $A, B, C, \ldots M, N, P, \ldots$ which describe the physical phenomena. As known, dimensional homogeneity is the *sine qua non* condition for the validity of rational relationships. As a consequence we can say that dimensional analysis offers a means for checking the validity of rational relationships — derived by other methods — for their compliance with the dimensional homogeneity conditions.

But dimensional homogeneity should also be respected in the physical relationships where logarithms are implied.

Dimensional analysis cannot be applied to rheology, since it helps neither to check the dimensional homogeneity nor to establish the general form of a physical relationship. Palacios [1.32][12.2] calls the rheological variables pseudo-physical quantities.

12.2 List of Physical Quantities

Specialists are unanimous in saying that, although important, the problem of drawing up a list of variables to describe a certain well-determined physical phenomenon is still a difficult task. As yet, no definite rules are known for such an undertaking and consequently we have no reliable basis for any statement concerning the accuracy of such a list, no matter how it was drawn up. Actually, such a list will primarily rely on the theoretical knowledge we have about the phenomenon under consideration, but also on the intuition of the individual who used the dimensional analysis. Often the following question arises : how can we know that some variables are influential in the case of a given physical phenomenon whilst others are not?

The description of physical phenomena is continuously developing and improved methods are devised as new data are acquired. However, it is well known that even physical phenomena which at a first seem simple may be difficult to describe. Thus, for instance, as Dragoș [3.4][12.3] points out, we do not know for sure what motion is, nor if it is a continuous or a discontinuous positional shift of one body with respect to other bodies. Newton thought of motion as a continuous change with time of the mobile co-ordinates. But he first had to devise a mathematical analysis to characterize motion and motion change phenomena. Without an analysis of continuous variations, the concepts of velocity and acceleration — which are fundamental features of motion — could not have been defined. This is why in order to list the physical quantities — all one can do is to produce a few general rules which aid, yet, do not provide the solution.

Rule I : a theory should be devised for the phenomenon under consideration so that a solution could be reached if the difficulties

[12.2] p. 36.
[12.3] p. 90.

raised by mathematical calculations could be overcome. Consequently, a physical model should be first built up for the phenomenon to explain why and how the variables affect the said phenomenon; even a raw model can evidence the characteristic variables. This rule relies on the fact that dimensional analysis is subordinated to physical theories and not the other way around. If this rule is overlooked, situations could arise where dimensional analysis could lead to accurate solutions for absurd problems.

As an illustration let us consider the problem put by Selwyn[12.4]: what is the training time t needed for a golf player to hit the ball a distance l? Since it is evident that we have to pay due consideration to the normal acceleration of free fall g, we shall reach the conclusion that the function of variables

$$f(t, l, g) \tag{12.12}$$

is identical to that of the simple pendulum after the elimination of the oscillation amplitude. As a consequence, the accurate solution will be

$$t = k\sqrt{\frac{l}{g}} \tag{12.13}$$

where k is a numerical constant.

Rule II: the list of variables describing a physical phenomenon should include only independent variables, except for the dependent one. That is to say that the variables considered when describing a well-determined physical phenomenon should be connected through one single relationship.

Consider, for example, the vertical fall of a body under the weight of its own mass, m. The only force involved will be its own weight G. But the evolution of the phenomenon also involves time t, velocity v, acceleration a and the space s. Variables m, G and t are independent while the rest are dependent. The implicit function of $m^* = 6$ variables will therefore be

$$f(m, G, T, v, a, s) = 0. \tag{12.14}$$

12.4 Selwyn; in *Brit. J. Appl. Physics*, IV, 1935, p. 202.

The dimensional equations of the variables for $d^* = 3$ will result from the following dimensional matrix:

$$\begin{array}{c|cccccc|} & m & G & t & v & a & s \\ \hline L & 0 & 1 & 0 & 1 & 1 & 1 \\ M & 1 & 1 & 0 & 0 & 0 & 0 \\ T & 0 & -2 & 1 & -1 & -2 & 0 \end{array} \tag{12.15}$$

whose rank is $r^* = 3$.

The number of non-dimensional products of powers will therefore be: $i^* = m^* - r^* = 3$.

According to Rayleigh's method, the relationship existing between the variables is written

$$t = km^b G^c v^d a^e s^h, \tag{12.16}$$

then the dimensional equations given by matrix (12.15) are introduced in the above relation. The dimensional homogeneity condition is expressed by the system of linear, non-homogeneous equations:

$$\begin{aligned} c + d + e + h &= 0 \qquad \text{(L)} \\ b + c &= 0 \qquad \text{(M)} \\ -2c - d - 2e &= 1 \qquad \text{(T)} \end{aligned} \tag{12.17}$$

which, for $e^* = 3 < n^* = 6$ is an indeterminate system. It admits, say, the solution:

$$b = -c, \quad d = -1 - 2c - 2e, \quad h = 1 + c + e. \tag{12.18}$$

Hence, for this solution, we can write the following matrix:

$$\begin{array}{c|cccccc|} & b & c & & d & e & h \\ & m & G & t & v & a & s \\ \hline \Pi_1 & 0 & 0 & 1 & 1 & 0 & -1 \\ \Pi_2 & -1 & 1 & 0 & -2 & 0 & 1 \\ \Pi_3 & 0 & 0 & 0 & -2 & 1 & 1 \end{array} \tag{12.19}$$

And relation (12.16) becomes

$$t = k \frac{s}{v} \varphi \left(\frac{Gs}{mv^2}, \frac{as}{v^2} \right) \tag{12.20}$$

which is the solution of the problem.

However, the form of function φ cannot be established by dimensional analysis. The indeterminate results obtained from calculations by dimensional analysis is due to the fact that in function f all 6 variables were taken into consideration while only 3 were independent. Applying dimensional analysis to 3 different functions, including all independent variables plus a dependent one, leads to relations where non-determinate φ functions do not appear. Thus, function

$$f(m, G, t, v) = 0 \tag{12.21}$$

leads to relation

$$v = k \frac{Gt}{m}, \tag{12.22}$$

function

$$f(m, G, t, a) = 0 \tag{12.23}$$

leads to relation

$$a = k \frac{G}{m}, \tag{12.24}$$

while function

$$f(m, G, t, s) = 0 \tag{12.25}$$

leads to relation

$$s = k \frac{Gt^2}{m}. \tag{12.26}$$

Rule III : all universal and characteristic constants taking part in the evolution of a physical phenomenon should be taken into consideration according to the theory of the process. Obviously, there can be no satisfactory solution for a problem, of say, elasticity, if we fail to take into consideration all elasticity constants involved in the evolution of the relevant phenomenon. This need to pay due consideration to all constants raises important difficulties when drawing the list of variables which are characteristic of a given physical phenomenon.

With only one constant omitted the computations conducted through dimensional analysis could lead to absurd results. This is the conclusion reached when considering Labocetta's problem [12.5]. The statement of this problem is as follows : it is required to demonstrate that if the densities of the bodies which build up the solar system are kept constant while the lengths are multiplied by a constant factor, the planets' periods of revolution would be the same. Hence, this is a problem where only inertial forces are involved ; as a consequence, periods t should be a function of planet masses, system configurations and gravitational constant. On the other hand, planet masses could be expressd by the product of density ρ by the corresponding volume, which is a function of a length measure given by the radius. Therefore, the system configuration will be described by a number of radii r, distances l and angles α, i.e. lengths and constant numbers. The gravitational constant is involved in the universal attraction law — which is expressed by the known relation (2.8) ; it is denoted by k'. In what follows, for the sake of simplification, only two bodies will be considered ; the conclusions will, however, have a general validity. The list of variables involved in this problem leads to the following implicit function

$$f(t, \rho_1, \rho_2, r_1, r_2, l_1, l_2, k') = 0, \tag{12.27}$$

where angles α were not included since they have no dimensional equation.

For $d^* = 3$, the dimensional equations of variables result from the following dimensional matrix :

	t	ρ_1	ρ_2	r_1	r_2	l_1	l_2	k'
L	0	−3	−3	1	1	1	1	3
M	0	1	1	0	0	0	0	−1
T	1	0	0	0	0	0	0	−2

(12.28)

The rank of the dimensional matrix (12.28) is $r^* = 3$.

[12.5] Labocetta, *Electrotechnica*, 1832, p. 1629.

Hence, the number of non-dimensional products of powers $i^* = 8 - 3 = 5$.

In compliance with Rayleigh's method, the relation connecting the $m^* = 8$ variables is written

$$t = k\,\rho_1^a\,\rho_2^b r_1^c r_2^d l_1^e l_2^h k'^i. \tag{12.29}$$

Now, in relation (12.29) we introduce the dimensional equations given by the dimensional matrix (12.28). The condition of dimensional homogeneity is expressed by the system of non-homogeneous linear equations:

$$\begin{aligned} -3a - 3b + c + d + e + h + 3i &= 0 \quad &(L)\\ a + b + \ \ i &= 0 \quad &(M)\\ -2i &= 1 \quad &(T). \end{aligned} \tag{12.30}$$

The above indeterminate system of equations admits the following solution:

$$a = -\frac{1}{2} - b, \quad c = -d - e - h, \quad i = -\frac{1}{2} \tag{12.31}$$

Hence, the solution matrix is:

$$\begin{array}{c|cccccccc|} & & a & b & c & d & e & h & i \\ & t & \rho_1 & \rho_2 & r_1 & r_2 & l_1 & l_2 & k' \\ \hline \Pi_1 & 1 & \frac{1}{2} & 0 & 0 & 0 & 0 & 0 & \frac{1}{2} \\ \Pi_2 & 0 & -1 & 1 & 0 & 0 & 0 & 0 & 0 \\ \Pi_3 & 0 & 0 & 0 & -1 & 1 & 0 & 0 & 0 \\ \Pi_4 & 0 & 0 & 0 & -1 & 0 & 1 & 0 & 0 \\ \Pi_5 & 0 & 0 & 0 & -1 & 0 & 0 & 1 & 0 \end{array} \tag{12.32}$$

and as a consequence, relation (12.29) becomes

$$t = k\frac{1}{\sqrt{\rho_1 k'}}\varphi\left(\frac{\rho_2}{\rho_1}, \frac{r_2}{r_1}, \frac{l_1}{r_1}, \frac{l_2}{r_1}\right), \tag{12.33}$$

where function φ is undetermined from dimensional analysis.

Assuming that the list of variables omits the gravitational constant, the dimensional analysis computations will lead to the elimination of variable t as well, and relation (12.33) will take the form :

$$\frac{\rho_2}{\rho_1} = k\varphi\left(\frac{r_2}{r_1},\ \frac{l_1}{r_1},\ \frac{l_2}{r_1}\right) \tag{12.34}$$

which is, of course, absurd since it states that the ratio of specific masses of two bodies would be a function of certain given factors of shape.

12.3 Number of Fundamental Quantities

Dimensional analysis computations are carried out on the basis of dimensional equations of variables, which are physical quantities and universal or characteristic constants. Dimensional equation are expressed as functions of fundamental quantities of a coherent system of units. As known, we are now using systems with 3 to 4 fundamental units, therefore having the same number of fundamental quantities. As known also, systems with a greater number of fundamental units as well as systems with no fundamental units at all were suggested (Section I.2.4).

Whenever a physical phenomenon is studied through dimensional analysis, the optimal number of fundamental quantities should always be used. If this rule is not observed, the result will be impaired, or even completely erratic. The applications given in this work obey this rule and the optimal number of fundamntal quantities is used. For those cases where a lower or a higher number of fundamental quantities is used, the following examples are given.

Obviously, by reducing the number of fundamental quantities d^*, the dimensional equations written for the variables become simpler. As a consequence, the condition of dimensional homogeneity of rational relations will be expressed by a smaller number of equations written for the dimension and the quantitiy exponents since, as established,

$$e^* \leqslant d^* ; \tag{10.4}$$

while this reduction takes place, the number of unknowns n^* is, however, kept constant, since, as established,

$$n^* = m^* - 1 \tag{10.5}$$

where m^* has the significance already given. The number of non-dimensional products of powers is correspondingly reduced. In this situation, the homogeneity condition is less effective and the dimensional analysis could lead to erroneous solutions as a result of the development of false non-dimensional products of powers.

To illustrate let us return to the problem of the simple pendulum for which

$$t = k\, l^a m^b G^c \varphi\,(\alpha). \tag{12.5}$$

In the system of units with $d^* = 2$ fundamental quantities, the dimensional equations of the variables will result from the following dimensional matrix :

$$\begin{array}{c|cccc|} & l & m & t & G \\ \hline \mathrm{L} & 1 & 3 & 0 & 4 \\ \mathrm{T} & 0 & -2 & 1 & -4 \end{array} \tag{12.35}$$

whose rank is $r^* = 2$.

Consequently, the number of non-dimensional products of powers $i^* = 2$.

The condition of dimensional homogeneity of relation (12.5) is expressed by the system of non-homogeneous linear equations :

$$\begin{aligned} a + 3b + 4c &= 1 \qquad (\mathrm{L}) \\ -\,2b + 4c &= 1 \qquad (\mathrm{T}) \end{aligned} \tag{12.36}$$

which admits, say, the solution :

$$b = 1 - a, \quad c = -\frac{3}{4} + \frac{1}{2}\, a \cdot \tag{12.37}$$

Therefore the solution matrix will be:

	a	b		c
	l	m	t	G
Π_1	0	−1	1	$\frac{3}{4}$
Π_2	1	−1	0	$\frac{1}{2}$

(12.38)

and relation (12.5) will take the form

$$t = k\,\frac{m}{G^{3/4}}\,\varphi\left(\frac{lG^{1/2}}{m},\ \alpha\right). \tag{12.39}$$

This should be compared to relation (12.10) derived before by using the system of units with $d^* = 3$ fundamental quantities (Section III.12.1); to facilitate this comparison, we consider a new non-dimensional product of powers:

$$\Pi_1' = \Pi_1 \Pi_2^{-1/2} = t\sqrt{\frac{g}{1}} \tag{12.40}$$

and finally we obtain

$$t = k\sqrt{\frac{1}{g}}\,\varphi\left(1\sqrt{\frac{g}{m}},\ \alpha\right). \tag{12.41}$$

As can be seen, apart from the non-dimensional products of powers, included in the corrrect relationship, relation (12.41) also includes the product Π_2; it follows that solution (12.41) is less accurate than relation (12.10) since the unknown function depends on two parameters instead of one. Furthermore, the non-dimensional product of powers Π_2 is erroneous since it attests that mass affects the pendulum period of oscillation, which is neither theoretically confirmed, nor experiementally checked.

It is also obvious that the increase in the number of fundamental quantities d^* increases the complexity of dimensional equations written for the variables. As a consequence, the condition of the dimensional homogeneity of physical relations will be expressed by a greater number of equations between the dimensional and quantity exponents ; at the same time, the unknowns will be the same. The number of non-dimensional products of powers should generally drop, but it may stay unchanged.

Let us return, for illustration purposes, to the problem of the simple pendulum, for which the function of variables

$$f\,(l, m, \alpha, t, G) = 0 \tag{12.3}$$

was used. The system of units with $d^* = 4$ fundamental quantities, which could be considered, should include, in addition to fundamental quantities L, M, T, the fundamental quantity F as well. The system of units with four fundamental quantities can be used only if, in any function of variables that describes a dynamic physical phenomenon, the dynamic constant will be introduced ; this constant is included in the fundamental law of dynamics — Newton's law — which is expressed by the already known relation (12.5). To avoid repetition, in this example, the dynamic constant will be denoted by k_1. As a consequence, the function of variables (12.3) should be added to include the dynamic constant ; thus

$$f\,(l, m, \alpha\,, t, G, k_1) = 0. \tag{12.42}$$

In the system of units chosen, the dimensional equations of variables will result from the following dimensional matrix :

$$\begin{array}{c|ccccc|} & l & m & t & G & k_1 \\ \hline \mathrm{L} & 1 & 0 & 0 & 0 & -1 \\ \mathrm{M} & 0 & 1 & 0 & 0 & -1 \\ \mathrm{T} & 0 & 0 & 1 & 0 & 2 \\ \mathrm{F} & 0 & 0 & 0 & 1 & 1 \end{array} \tag{12.43}$$

whose rank is $f^* = 4$.

In this case, the number of non-dimensional products of powers will be equal to unity : $i^* = 1$.

Relation (12.42), written according to Rayleigh's method, will have the form

$$t = kl^a m^b G^c k_1^d \varphi(\alpha). \tag{12.44}$$

For the above relation, the condition of dimensional homogeneity will be expressed by the system of non-homogeneous linear equations :

$$\begin{aligned} a - d &= 0 \quad &(L) \\ b - d &= 0 \quad &(M) \\ 2d &= 1 \quad &(T) \\ c + d &= 0 \quad &(F) \end{aligned} \tag{12.45}$$

which admits the solution :

$$a = \frac{1}{2}, \quad b = \frac{1}{2}, \quad c = -\frac{1}{2}, \quad d = \frac{1}{2}, \tag{12.46}$$

and the matrix of the solution will be :

$$\begin{array}{c|ccccc|} & a & b & & c & d \\ & l & m & t & G & k_1 \\ \hline \Pi & -\frac{1}{2} & -\frac{1}{2} & 1 & \frac{1}{2} & -\frac{1}{2} \end{array} \tag{12.47}$$

On the basis of relation (12.2) and matrix (12.47), the solution of the problem can be written in the form

$$t = k\sqrt{\frac{lk_1}{g}}\,\varphi(\alpha). \tag{12.48}$$

As can be seen when compared against the correct relation (12.10), relation (12.48) includes in addition the constant k_1. If we make the arbitrary choice $k_1 = 1$, then the system of units will be reduced to a system with three fundamental quantities and we shall find again relation (12.10), from relation (12.48).

Considering the possibility of adopting a system of units with various numbers of fundamental quantities, the following statement holds true : *whenever the increase in the original number of fundamental quantities results in a reduced number of independent non-dimensional products of powers, it follows that the original number is sufficient.* The consequences of any reduction in the number of fundamental quantities are therefore obvious.

The increase in the number of fundamental quantities so as to obtain more complex dimensional equations can also be produced by applying Huntley's method. According to this method, the dimensional equations of the variables will be written on the basis of oriented fundamental quantities. As shown before (Section I.5.3), this method consists of replacing the fundamental quantity of length L by three components, L_x, L_y, L_z, considered separately. The obvious effect of this separate consideration is the increase in the number of equations expressing the condition of dimensional homogeneity, which further leads to the decrease in the number of independent non-dimensional products of powers and, consequently, the solution will gain in accuracy.

To illustrate, let us consider the following problem : a projectile of mass m is horizontally launched at the initial velocity v at a height h which is maintained constant; what will be the distance x to the projectile's point of fall? To complete the description of the physical phenomenon, we shall also say that weight $G = mg$ will be implied.

The implicit function of the $m^* = 5$ variables involved in this physical phenomenon will therefore be

$$f(m, v, h, x, G) = 0. \tag{12.49}$$

The dimensional matrix, corresponding to the dimensional equations written for the variables in the system of units with $d^* = 3$ fundamental quantities, is :

	m	v	h	x	G
L	0	1	1	1	1
M	1	0	0	0	1
T	0	−1	0	0	−2

(12.50)

whose rank is $r^* = 3$.

In this case, the number of non-dimensional products of powers will be $i^* = 2$.

The explicit relation written for the dependent variable x according to Rayleigh's method is of the form

$$x = k\, m^a v^b h^c G^d. \tag{12.51}$$

Hence, the system of linear equations expressing the dimensional homogeneity condition for relation (12.51) is:

$$\begin{aligned} b + c + d &= 1 \qquad (L) \\ a + d &= 0 \qquad (M) \\ -b - 2d &= 0 \qquad (T) \end{aligned} \tag{12.52}$$

which is built out of $e^* = 3$ equations with $n^* = 4$ unknowns.

Assuming the solution of the indeterminate system of equations (12.52) to be:

$$a = -d, \quad b = -2d, \quad c = 1 + d \tag{12.53}$$

then the matrix of the solution will be easily written

(12.54)

	a m	b v	c h	x	d G
Π_1	0	0	−1	1	0
Π_2	−1	−2	1	0	1

Hence the solution of the problem will have a relation of the form

$$x = kh\varphi\left(\frac{hG}{mv^2}\right)^d \tag{12.55}$$

or, bearing in mind relation (12.2):

$$x = kh\varphi\left(\frac{hg}{v^2}\right)^d \tag{12.56}$$

where function φ is still unknown.

If the same physical phenomenon is treated by Huntley's method, the dimensional matrix will be

$$\begin{array}{c|ccccc|} & m & v & h & x & G \\ \hline L_x & 0 & 1 & 0 & 1 & 0 \\ L_z & 0 & 0 & 1 & 0 & 1 \\ M & 1 & 0 & 0 & 0 & 1 \\ T & 0 & -1 & 0 & 0 & -2 \end{array} \qquad (12.57)$$

whose rank is $r^* = 4$.

The number of non-dimensional products of powers will be $i^* = 1$.

The system of linear equations relating the dimensional to the quantity exponents and expressing the condition for the dimensional homogeneity has the form :

$$\begin{aligned} b &= 1 && (L_x) \\ c + d &= 0 && (L_z) \\ -a + d &= 0 && (M) \\ -a - 2d &= 0 && (T). \end{aligned} \qquad (12.58)$$

Hence, the solution of the determinate system of equations (12.58) is :

$$a = \frac{1}{2}, \quad b = 1, \quad c = \frac{1}{2}, \quad d = -\frac{1}{2}. \qquad (12.59)$$

The matrix of the solution is of the form :

$$\begin{array}{c|ccccc|} & a & b & c & & d \\ & m & v & h & x & G \\ \hline \Pi & -\frac{1}{2} & -1 & -\frac{1}{2} & 1 & \frac{1}{2} \end{array} \qquad (12.60)$$

On the basis of relation (12.2) and matrix (12.60), the solution of the problem will be expressed by the relation

$$x = kv\frac{h}{g}. \qquad (12.61)$$

As can be seen, relation (12.56) is less accurate than relation (12.61). For $d = -\frac{1}{2}$, from (12.56), relation (12.61) results.

13 Computing Non-dimensional Products of Powers

13.1 Application of the Π Theorem

Any function of m^* variables (physical quantities, universal and characteristic constants) can be reduced down to a function of i^* non-dimensional products of powers. The Π theorem or Rayleigh's method can be applied to this end. However, whenever a certain physical phenomenon is studied, only one of the above two methods should be applied, inasmuch as they both lead to the same result. As a general rule the Π theorem is recommended for functions where a comparatively greater number of variables are involved while Rayleigh's method will be reserved for problems with a smaller number of variables. This recommendation is based on the readiness with which non-dimensional products of powers can be derived.

To illustrate, let us consider the following physical phenomenon: a sphere of a diameter d and density ρ_0, falls under the action of its own weight G at a point on the earth where the normal acceleration of the free fall is g; the sphere is immersed in a viscous fluid of density ρ and dynamical viscosity η. Question: What will be the relation for the velocity v with which the sphere will fall when the motion becomes uniform? The apparent weight of the sphere G_0 is given by relation

$$G = \frac{1}{6} \pi d^3(\rho_0 - \rho)g. \tag{13.1}$$

When the acceleration of the motion is cancelled out, the sphere mass will no longer affect the fall; one should however take into consideration the density ρ of the fluid. From the physical model presented, the implicit function of the $m^* = 5$ variables will be written as

$$f(d, G, \rho, \eta, v) = 0. \tag{13.2}$$

The implicit function of variables taken into consideration in the following dimensional analysis computations is the one where the sphere weight is substituted by the normal acceleration of the free fall, that is

$$f(d, g, \rho, \eta, v) = 0. \tag{13.3}$$

Referring to the system of units with $d^* = 3$ fundamental quantities, the dimensional matrix of variables is of the form:

$$\begin{array}{c|ccccc|} & d & g & \rho & \eta & v \\ \hline L & 1 & 1 & -3 & -1 & 1 \\ M & 0 & 0 & 1 & 1 & 0 \\ T & 0 & -2 & 0 & -1 & -1 \end{array} \tag{13.4}$$

and has the rank $r^* = 3$.

The number of non-dimensional independent products of powers of a complete series will be $i^* = 2$. Bearing in mind that the independent variable is velocity v, the following four groups of quantities considered to be fundamental can be built up:

$$\begin{aligned} &\text{group 1, quantities: } g, \rho, \eta \\ &\text{group 2, quantities: } d, \rho, \eta \\ &\text{group 3, quantities: } d, g, \eta \\ &\text{group 4, quantities: } d, g, \rho. \end{aligned} \tag{13.5}$$

Relations written for the non-dimensional products of powers according to the Π theorem are, for the 1st group:

$$\Pi_1' = \frac{d}{g^a \rho^b \eta^c}, \quad \Pi_1'' = \frac{v}{g^a \rho^b \eta^c}. \tag{13.6}$$

The systems of linear equations relating the dimensional to the quantity exponents and expressing the condition of dimensional homogeneity are readily derived :

$$1' \begin{cases} a - 3b - c = 1 \\ b + c = 0 \\ -2a - c = 0 \end{cases} \qquad 1'' \begin{cases} a - 3b - c = 1 & \text{(L)} \\ b + c = 0 & \text{(M)} \\ -2a - c = -1 & \text{(T).} \end{cases} \tag{13.7}$$

The solution of the determinate systems of linear equations (13.7) expressing the condition of dimensional homogeneity are :

$$1' \begin{cases} a = -\dfrac{1}{2} \\ b = -\dfrac{2}{3} \\ c = \dfrac{2}{3} \end{cases} \qquad 1'' \begin{cases} a = \dfrac{1}{3} \\ b = -\dfrac{1}{3} \\ c = \dfrac{1}{3} \end{cases} \tag{13.8}$$

while the solution matrix is :

	d	g (a)	ρ (b)	η (c)	v
Π'_1	1	$\dfrac{1}{3}$	$\dfrac{2}{3}$	$-\dfrac{2}{3}$	0
Π''_1	0	$-\dfrac{1}{3}$	$\dfrac{1}{3}$	$-\dfrac{1}{3}$	1

(13.9)

The relation, as expressed by the physical quantities and the non-dimensional product of powers, is

$$v = k \frac{g^{1/3} \eta^{1/3}}{\rho^{1/3}} \varphi \left(\frac{d g^{1/2} \rho^{2/3}}{\eta^{2/3}} \right). \tag{13.10}$$

For the 2nd group, the relations of non-dimensional products of powers are :

$$\Pi'_2 = \frac{g}{d^a \rho^b \eta^c}, \quad \Pi''_2 = \frac{v}{d^a \rho^b \eta^c} \tag{13.11}$$

while the systems of linear equations expressing the condition of dimensional homogeneity will be of the form :

$$2'\begin{cases} a - 3b - c = 1 \\ b + c = 0 \\ -c = -2 \end{cases} \quad 2''\begin{cases} a - 3b - c = 1 & \text{(L)} \\ b + c = 0 & \text{(M)} \\ -c = -1 & \text{(T)} \end{cases} \tag{13.12}$$

wherefrom the solutions of the determinate systems of linear equations (13.12) are readily found :

$$2'\begin{cases} a = -3 \\ b = -2 \\ c = 2 \end{cases} \quad 2''\begin{cases} a = -1 \\ b = -1 \\ c = 1. \end{cases} \tag{13.13}$$

Hence the solution matrix is :

	a	b	c		
	d	g	ρ	η	v
Π'_2	3	1	2	−2	0
Π''_2	1	0	1	−1	1

(13.14)

The relation, as expressed by the physical quantities and the non-dimensional product of powers, will be of the form

$$v = k\frac{\eta}{d\rho}\varphi\left(\frac{d^3 g\rho^2}{\eta^2}\right). \tag{13.15}$$

For the 3rd group the relation written for non-dimensional products of powers will have the form :

$$\Pi'_3 = \frac{\rho}{d^a g^b \eta^c}, \quad \Pi''_3 = \frac{v}{d^a g^b \eta^c} \tag{13.16}$$

while for the 4th group :

$$\Pi'_4 = \frac{\eta}{d^a g^b \rho^c}, \quad \Pi''_4 = \frac{v}{d^a g^b \rho^c}. \tag{13.17}$$

For the 3rd group, the relations as expressed by the corresponding physical quantities and non-dimensional products of powers will be of the form

$$v = k\sqrt{dg}\;\varphi\left(\frac{d^{3/2}g^{1/2}\rho}{\eta}\right) \tag{13.18}$$

while for the 4th group :

$$v = k\sqrt{dg}\;\varphi\left(\frac{\eta}{d^{3/2}g^{1/2}\rho}\right). \tag{13.19}$$

13.2 Application of Rayleigh's Method

In what follows, we shall describe the dimensional analysis computations as conducted according to Rayleigh's method.

The explicit relation written for the dependent variable will be of the form

$$v = k\, d^a g^b \rho^c \eta^e \tag{13.20}$$

and the system of linear equations expressing the condition of dimensional homogeneity can be readily written :

$$\begin{aligned} a + b - 3c &= 1 \quad \text{(L)} \\ c + e &= 0 \quad \text{(M)} \\ -2b - e &= -1 \quad \text{(T)}. \end{aligned} \tag{13.21}$$

As seen, the system of equation (13.21) is indeterminate since the number of equations is $e^* = 3$ while the number of unknowns, $n^* = 4$; to solve such a system an unknown should be chosen at random ; as a consequence, four solutions could be found. The solution of the system of linear equations (13.21) – indeterminate for the random unknown a – is :

$$b = \frac{1}{3} + \frac{1}{3}a, \quad c = -\frac{1}{3} + \frac{2}{3}a, \quad e = \frac{1}{3} - \frac{2}{3}a \tag{13.22}$$

while the matrix of the solution is :

	a d	b g	c ρ	e η	v
Π_1	1	1/3	1/3	−1/3	1
Π_2	1	−1/3	2/3	−2/3	0

(13.23)

The relation, as expressed by the physical quantities and the non-dimensional product of powers, will be

$$v = k\,\frac{g^{1/3}\eta^{1/3}}{\rho^{1/3}}\,\varphi\left(\frac{d\,g^{1/3}\rho^{2/3}}{\eta^{2/3}}\right). \tag{13.10}$$

Now the solution for the system of linear equations (13.21), which is indeterminate for random unknowns b, c and e, are obvious.

As seen, the Π theorem and Rayleigh's method have both led to the same relations (13.10), (13.15), (13.18) and (13.19).

★

As established, any non-dimensional product of powers where only two quantities of the like nature (hence quantities having the same dimensional equation) are involved is called a factor of form. When functions of variables include two or several quantities of the like nature, a lesser amount of dimensional analysis computations are required to find the non-dimensional products of powers.

14 Relationships Between Physical Quantities

14.1 Stellar Vibrations

An approach to this problem by dimensional analysis was suggested by Rayleigh. Any star is a fluid body, whose compactness is maintained by its own gravity and consequently a star can vibrate in a variety of ways. Of these kinds of vibration the harmonic in-phase vibrations are particularly important in theory since the particles are passing simultaneously through the neutral position and, still simultaneously, reach the extreme positions. Such a vibration is called the natural mode of vibration. The simplest natural mode of vibration for a star is that motion whereby its surface alternately takes the shape of a flattened and of an elongated ellipsoid; however, both shapes are symmetrical with respect to two fixed axes. If the system does not incur frictions, the superposition of several natural modes of vibration will result in a small vibration.

Any vibration takes place at a certain frequency ν. As already known, a low viscosity will not substantially affect the system frequency and hence this variable can be neglected. Furthermore, for the sake of simplicity, we shall assume that the star density ρ is constant. Therefore, the variables likely to affect the frequency of a vibration knot will be: star diameter D, density ρ and the gravitational constant k'. On the basis of this physical model, computations of dimensional analysis can be carried out [1.26] [14.1].

The implicit function of $m^* = 4$ variables is

$$f(\nu, D, \rho, k') = 0 \tag{14.1}$$

while the dimensional matrix of variables for $d^* = 3$ is:

(14.2)

	ν	D	ρ	k'
L	0	1	−3	3
M	0	0	1	−1
T	−1	0	0	−2

whose rank is $r^* = 3$.

The number of non-dimensional products of powers $i^* = 1$.

The explicit relation of the dependent variable is:

$$\nu = k\, D^a \rho^b k'^c. \tag{14.3}$$

The system of linear equations expressing the condition of dimensional homogeneity is:

$$\begin{aligned} a - 3b + 3c &= 0 \quad \text{(L)} \\ b - c &= 0 \quad \text{(M)} \\ -2c &= -1 \quad \text{(T)} \end{aligned} \tag{14.4}$$

[14.1] p. 32.

hence the solution for the determinate system of equations is :

$$a = 0, \quad b = \frac{1}{2}, \quad c = \frac{1}{2} \tag{14.5}$$

and the solution matrix is :

$$\begin{array}{c|cccc|} & & a & b & c \\ & \nu & D & \rho & k' \\ \hline \Pi & 1 & 0 & -\frac{1}{2} & -\frac{1}{2} \end{array} \tag{14.6}$$

The relation, as expressed through physical quantities, is

$$\nu = k\sqrt{\rho k'}. \tag{14.7}$$

It was therefore demonstrated that the frequency of any natural mode of stellar vibration is independent of the diameter and in direct proportion to the second-order root of the density. This conclusion is valid for variable density stars as well, provided it can be assumed that the latter have the same mass distribution; in such a case, density ρ may be, say, that of the star centre.

14.2 Effects of Temperature on Gas Viscosity

In 1899, Rayleigh applied, in a very ingenious way, dimensional analysis to the study of effects induced by temperature on gas viscosity [14.2, 14.3] [1.26]. There are many applications of the gas kinetic theory where it is not necessary to take into consideration molecule structure. The molecule could be compared to a billiard ball. The attraction forces between molecules can be neglected. Whenever two molecules are getting close to one another, a repulsion force is developed by each molecule until they are close to contact. This force is proportional to a power of the distance, i.e.

$$F = Kd^n \tag{14.8}$$

[14.2] Rayleigh, *Proc. Roy. Soc.* London, LXVI, 1899–1900, p. 68–74.
[14.3] Rayleigh, *Nature* LXXXXV, 1915.

where d is the distance between molecule centres while n is a numerical exponent. The numerical factor of proportionality K is a characteristic of the molecule. From relation (14.8), the following dimensional equation results :

$$[K] = L^{1+n} M T^{-2}. \tag{14.9}$$

Approaching this problem, Rayleigh considered that viscosity is independent of density; this conclusion was drawn by Maxwell from the study of the molecules. It was experimentally shown that this law of viscosity independence of density holds true for pressures varying from 20 mm Hg up to one atmosphere and that it is not valid for pressures in excess of a few atmospheres, owing to the intermolecular forces involved in the case of high density gases. Of course, if the viscosity of a gas is independent of density, it will also be independent of the molecular characteristics which control the density, i.e. the number of molecules per unit volume and the molecular free path. As a consequence, viscosity can be determined by the molecule mass m, by the molecule mean velocity v and by the repulsion coefficient K.

The implicit function of ($m^* = 4$) variables is :

$$f(\eta, m, v, K) = 0 \tag{14.10}$$

and the dimensional matrix for $d^* = 3$ is :

$$\begin{array}{c|cccc|} & \eta & m & v & K \\ \hline \mathrm{L} & -1 & 0 & 1 & n+1 \\ \mathrm{M} & 1 & 1 & 0 & 1 \\ \mathrm{T} & -1 & 0 & -1 & -2 \end{array} \tag{14.11}$$

whose rank is $r^* = 3$.

The number of non-dimensional products of powers $i^* = 1$.

The explicit relation of the dependent variable is

$$\eta = k m^a v^b K^c \tag{14.12}$$

while the system of linear equations expressing the condition of dimensional homogeneity is :

$$\begin{aligned} b + (n+1)c &= -1 \quad &(\mathrm{L}) \\ a + c &= 1 \quad &(\mathrm{M}) \\ -b - 2c &= -1 \quad &(\mathrm{T}). \end{aligned} \tag{14.13}$$

The solution for the determinate system of equations is :

$$a = \frac{n+1}{n-1}, \quad b = \frac{n+3}{n-1}, \quad c = \frac{-2}{n-1} \tag{14.14}$$

while the solution matrix is :

	η	a m	b v	c K
Π	1	$\frac{n+1}{n-1}$	$-\frac{n+3}{n-1}$	$\frac{2}{n-1}$

(14.15)

The relation, as expressed by physical quantities, is

$$\eta = k m^{\frac{n+1}{n-1}} v^{\frac{n+3}{n-1}} K^{\frac{-2}{n-1}}. \tag{14.16}$$

In the kinetic theory of gases, it is demonstrated that the absolute temperature θ of a gas is proportional to the kinetic energy of a molecule $\frac{1}{2}mv^2$. As a consequence, relation (14.16) can be written in the form :

$$\eta = k_1 m^{\frac{1}{2}} K^{\frac{-2}{n-1}} \theta^s \tag{14.17}$$

where

$$s = \frac{1}{2} + \frac{2}{n-1}, \tag{14.18}$$

while k_1 is a constant. For a certain given gas, m and K have constant values. Since $n > 1$, relations (14.17) and (14.18) will show that viscosity rises with temperature. Now, bearing in mind the fact that, according to relation (14.17), viscosity is expressed by a relation of the form

$$\eta = k_2 \, \theta^s \tag{14.19}$$

where k_2 is a new constant, we can see that, if represented on a system of logarithmic co-ordinates, this relation — which shows that viscosity is proportional to a power of temperature — will in fact be a straight line.

Relations (14.17) and (14.18) supply new information on the repulsion forces between molecules. For $n = 5$ we obtain $s = 1$, which shows that viscosity is proportional to the unit power of the absolute temperature; for $n = \infty$, we obtain $s = 0.5$, which shows that viscosity is proportional to the fourth root of the absolute temperature. Experimental measurements conducted by Rayleigh yield: for air, $s = 0.754$; for hydrogen, $s = 0.681$; for argon, $s = 0.815$; these results show that for common gases, n varies from 7 to 12 [1.26] [14.4].

14.3 Vibration of a Liquid Droplet

Rayleigh considers that the study of this problem is an example where dimensional analysis immediately leads to a solution that would be difficult to obtain by using the theory of this physical phenomenon [1.32] [14.5].

When at rest, the shape of the liquid droplet can be taken as a sphere of radius r. The liquid is characterized by density ρ. The spherical shape of the droplet is the result of forces generated by the superficial tension σ. Vibration period is t.

The implicit function of $m^* = 4$ variables is

$$f(r, \rho, \sigma, t) = 0 \tag{14.20}$$

while the dimensional matrix of variables for $d^* = 3$ is:

$$\begin{array}{c|cccc|} & r & \rho & \sigma & t \\ \hline \mathrm{L} & 1 & -3 & 0 & 0 \\ \mathrm{M} & 0 & 1 & 1 & 0 \\ \mathrm{T} & 0 & 0 & -2 & 1 \end{array} \tag{14.21}$$

whose rank is $r^* = 3$.

The number of non-dimensional products of powers $i^* = 1$.

The explicit relation of the dependent variable is

$$t = k\, r^a \rho^b \sigma^c. \tag{14.22}$$

[14.4] p. 54.
[14.5] p. 81.

The system of linear equations expressing the condition of dimensional homogeneity is:

$$
\begin{aligned}
a - 3b &= 0 \qquad (\mathrm{L}) \\
b + c &= 0 \qquad (\mathrm{M}) \\
-2c &= 1 \qquad (\mathrm{T}).
\end{aligned}
\tag{14.23}
$$

The solution of the determinate system of equations is:

$$a = \frac{3}{2}, \quad b = \frac{1}{2}, \quad c = -\frac{1}{2}. \tag{14.24}$$

The solution-matrix is:

$$
\begin{array}{c|cccc|}
 & a & b & c & \\
 & r & \rho & \sigma & t \\
\hline
\Pi & -\frac{3}{2} & -\frac{1}{2} & -\frac{1}{2} & 1
\end{array}
\tag{14.25}
$$

The relation, as expressed by the physical quantities, is

$$t = k\sqrt{\frac{r^3\rho}{\sigma}}. \tag{14.26}$$

It can be stated therefore that the square of the period is proportional to the cube of the radius and to the density, and inversely proportional to the superficial tension.

This problem may be given a general form by taking into consideration the gravitational action and by assuming that the droplet lies on an horizontal plane. Owing to symmetry the droplet surface will be a revolution surface and hence all that is needed is to consider radius r of the maximal horizontal cross-section. The implicit function of variables will therefore be

$$f(r, \rho, \sigma, t, \gamma) = 0 \tag{14.27}$$

where γ is the specific weight. In this case, dimensional analysis leads to a relation between two non-dimensional products of powers readily established.

14.4 Atom Oscillation in a Solid Body

Einstein suggested in 1907 a relation for the oscillation frequency ν of atoms within a solid [14.6]; he derived this relation through dimensional analysis. According to Einstein, the physical phenomenon considered here involved the following variables: atom mass m, distance between atoms d and solid compressibility μ (2.24).

The implicit function of $m^* = 4$ variables is

$$f(\nu, m, d, \mu) = 0. \tag{14.28}$$

The dimensional matrix of variables for $d^* = 3$ is:

	ν	m	d	μ
L	0	0	1	0
M	0	1	0	−1
T	−1	0	0	2

(14.29)

whose rank is $r^* = 3$.

The number of non-dimensional products of powers $i^* = 1$.

The explicit relation of the dependent variable is

$$\nu = k m^a d^b \mu^c. \tag{14.30}$$

The system of linear equations expressing the condition of dimensional homogeneity is:

$$\begin{aligned} a - c &= 0 \quad (\mathrm{L}) \\ b + c &= 0 \quad (\mathrm{M}) \\ 2c &= -1 \quad (\mathrm{T}). \end{aligned} \tag{14.31}$$

The solution of the determinate system of equations is:

$$a = -\frac{1}{2}, \quad b = \frac{1}{2}, \quad c = -\frac{1}{2}. \tag{14.32}$$

14.6 Einstein, A., *Annalen der Physik*, XXXV, 1911, p. 688.

The solution matrix is:

$$\begin{array}{c|cccc|} & & a & b & c \\ & \nu & m & d & \mu \\ \hline \Pi & 1 & \frac{1}{2} & -\frac{1}{2} & \frac{1}{2} \end{array} \tag{14.33}$$

The relation, as expressed by physical quantities, is

$$\nu = k\sqrt{\frac{d}{m\mu}}. \tag{14.34}$$

Einstein makes the additional assumption that the mass of the atom can be assimilated to that of a cube whose density is ρ, while the side is equal to the distance between the atoms d, i.e.

$$m = \rho d^3. \tag{14.35}$$

Relations (14.34) and (14.35) lead to the final relation:

$$\nu = k\frac{1}{\sqrt[6]{m^2\rho\mu^3}}. \tag{14.36}$$

By other means, it was found that the value of the proportionality factor is $k = 2.8 \times 10^2$. The validity of this relation was checked with good results for many bodies for which the atomic frequency was found by other methods.

When applied to the function of variables from relation (14.36), the dimensional analysis

$$f(\nu, m, \rho, \mu) = 0, \tag{14.37}$$

will lead to the correct relation.

15 Relationships between Non-dimensional Products of Powers

15.1 Intensity of Electric Current through a Transmission Line

Vaschy used dimensional analysis to study a physical phenomenon before even stating the Π theorem [1.35]. He considers a transmission line of a length l, capacity C and resistance R. At one end of the line, at time zero, an electromotive force E is applied. It is necessary to estimate the intensity of the electric current i supplied to the other end of the line at time t.

The implicit function of $m^* = 6$ variables is

$$f(l,\ C,\ R,\ E,\ i,\ t) = 0. \tag{15.1}$$

The dimensional matrix of variables for $d^* = 4$, length L, time T, energy W and voltage V, is:

$$\begin{array}{c|cccccc|} & l & C & R & E & i & t \\ \hline \mathrm{L} & 1 & 0 & 0 & 0 & 0 & 0 \\ \mathrm{T} & 0 & 0 & 1 & 0 & -1 & 1 \\ \mathrm{W} & 0 & 1 & -1 & 0 & 1 & 0 \\ \mathrm{V} & 0 & -2 & 2 & 1 & -1 & 0 \end{array} \tag{15.2}$$

whose rank $r^* = 4$.

Number of non-dimensional products of powers $i^* = 2$.

The explicit relation of the dependent variable is

$$i = kl^a C^b R^c E^d t^e. \tag{15.3}$$

The system of linear equations expressing the condition of dimensional homogeneity is:

$$\begin{aligned} a &= 0 && \text{(L)} \\ c + e &= -1 && \text{(T)} \\ b - c &= 1 && \text{(W)} \\ -2b + 2c + d &= -1 && \text{(V)} \end{aligned} \tag{15.4}$$

while the solution of the indeterminate system of equations for the random unknown b is:

$$a = 0,\ c = -1 + b,\ d = 1,\ e = -b \tag{15.5}$$

and the solution matrix is :

$$\begin{array}{c|cccccc|}
 & a & b & c & d & & e \\
 & l & C & R & E & i & t \\
\hline
\Pi_1 & 0 & 0 & 1 & -1 & 1 & 0 \\
\Pi_2 & 0 & 1 & 1 & 0 & 0 & -1
\end{array} \tag{15.6}$$

The relation, as expressed by the physical quantities and the non-dimensional product of powers, is

$$i = k\,\frac{E}{R}\,\varphi\left(\frac{CR}{t}\right). \tag{15.7}$$

It will be seen that function f with six variables was reduced to a function φ with only two variables ; however, dimensional analysis computation fails to yield the form of function φ. The result found by Vaschy agrees with the theory suggested by Kelvin.

However, the calculations overlooked the effects of self-induction; when this property is taken into consideration too, the function of variables (15.1) should be added to the line self-induction $\mathscr{L}$. The dimensional equation for this physical quantity is

$$[\mathscr{L}] = \mathrm{T}^2\mathrm{W}^{-1}\mathrm{V}^2. \tag{15.8}$$

Dimensional analysis computations which can be carried out in a similar manner for the function with $m^* = 7$ variables

$$f(l,\ C,\ R,\ E,\ i,\ t,\ \mathscr{L}) = 0 \tag{15.9}$$

lead to relation

$$i = k\,\frac{E}{R}\,\varphi\left(\frac{CR}{t},\quad \frac{CR^2}{\mathscr{L}}\right), \tag{15.10}$$

which is less accurate owing to the fact that the unknown function includes two independent variables. Compared to relation (15.7), relation (15.10) affords a better description for the relevant physical phenomenon; this relation should be used whenever the effects of self-induction become evident [1.35].

15.2 Cooling of a Solid Body Immersed in a Liquid Flow

The problem suggested by Boussinesq is a very much debated example of an application of dimensional analysis. A solid of a characteristic size l is immersed in a liquid flowing at a velocity v. It is maintained at a constant temperature θ, higher than the temperature θ_0 which is that of the liquid at a comparatively great distance from the solid body. The specific heat of the liquid as referred to the unit volume is c, while thermal conductivity is λ. It is necessary to estimate the amount of heat lost by the solid in the unit time q. The differential temperature will be denoted by $\Delta\theta = \theta - \theta_0$.

Rayleigh applied the dimensional analysis to Boussinesq's problem [15.1].

The implicit function of $m^* = 6$ variables is

$$f(l, v, \Delta\theta, c, \lambda, q) = 0. \qquad (15.11)$$

The dimensional matrix of variables for $d^* = 4$, length L, time T, temperature Θ and the amount of heat Q, is:

$$\begin{array}{c|cccccc|} & l & v & \Delta\theta & c & \lambda & q \\ \hline \mathrm{L} & 1 & 1 & 0 & -3 & -1 & 0 \\ \mathrm{T} & 0 & -1 & 0 & 0 & -1 & -1 \\ \Theta & 0 & 0 & 1 & -1 & -1 & 0 \\ \mathrm{Q} & 0 & 0 & 0 & 1 & 1 & 1 \end{array} \qquad (15.12)$$

whose rank is $r^* = 4$.

The number of non-dimensional products of powers $i^* = 2$.

The explicit relation of the dependent variable is

$$q = k\, l^a v^b \Delta\theta^d c^e \lambda^g. \qquad (15.13)$$

The system of linear equations expressing the condition of dimensional homogeneity is:

$$\begin{aligned} a + b - 3e - g &= 0 \ (\mathrm{L}) \\ d - e - g &= 0 \ (\mathrm{H}) \\ -b - g &= -1 \ (\mathrm{T}) \\ e + g &= 1 \ (\mathrm{Q}). \end{aligned} \qquad (15.14)$$

15.1 Raylelgh, *Nature*, LXXXXV, 1915 p. 66.

The solution of the indeterminate system of equations for the random unknown e is :

$$a = 1 + e,\ b = e,\ d = 1,\ g = 1 - e. \tag{15.15}$$

The solution matrix is :

	a l	b v	d $\Delta\theta$	e c	g λ	q
Π_1	−1	0	−1	0	−1	1
Π_2	1	1	0	1	−1	0

(15.16)

The relation, as expressed by the physical quantities and the non-dimensional product of powers, is

$$q = kl\Delta\theta\lambda\varphi\left(\frac{lvc}{\lambda}\right). \tag{15.17}$$

It is thus demonstrated that heat loss is proportional to the excess of temperature. If the size of the solid as well as the fluid velocity and nature are changed so that no changes will result in the value of the product of powers within the brackets, the heat loss will be proportional to the linear sizes of the solid and to the fluid thermal conductibility.

In the dimensional analysis computations conducted by Rayleigh, heat was considered to be independent of the mechanical quantities : this is explained by the fact that at that time the equivalence between mechanical work and heat had not been found.

Rayleigh's solution was criticized by Riabouchinsky [15.2], the latter considering the thermodynamic temperature to be a derived quantity which he called "molecule mean energy". Dimensional analysis computations conducted by Riabouchinsky rely on this assumption.

The dimensional matrix of variables for $d^* = 3$, length L, time T, heat amount Q, is

	l	v	$\Delta\theta$	c	λ	q
L	1	1	0	−3	−1	0
T	0	−1	0	0	−1	−1
Q	0	0	1	0	0	1

(15.18)

whose rank is $r^* = 3$.

15.2 Riabouchinsky, D., *Nature*, LXXXXV, 1915 p. 591.

The number of non-dimensional products of powers $i^* = 1$.
The system of linear equations expressing the condition of dimensional homogeneity is:

$$\begin{aligned} a + b - 3e - g &= 0 \quad (L) \\ -b - g &= -1 \quad (T) \\ d &= 1 \quad (Q). \end{aligned} \tag{15.19}$$

The solution of the indeterminate system of equations for the random unknowns b and e is:

$$a = 1 - 2b + 3e, \quad d = 1, \quad g = 1 - b. \tag{15.20}$$

The solution matrix is:

$$\begin{array}{c|cccccc|} & a & b & d & e & g & \\ & l & v & \Delta\theta & c & \lambda & q \\ \hline \Pi_1 & -1 & 0 & -1 & 0 & -1 & 1 \\ \Pi_2 & -2 & 1 & 0 & 0 & -1 & 0 \\ \Pi_3 & 3 & 0 & 0 & 1 & 0 & 0 \end{array} \tag{15.21}$$

The relation as expressed by the physical quantities and by the non-dimensional products of powers is

$$q = kl\Delta\theta\lambda\varphi\left(\frac{v}{l^2\lambda},\ l^3c\right). \tag{11.22}$$

Obviously, the result reached by Riabouchinsky is less accurate since function φ involves two non-dimensional products of powers. And yet, in his answer, Rayleigh could not find arguments against Riabouchinsky's reasoning. Indeed, his reply was: "The problem raised by Riabouchinsky is more close to the field of pure logic than to the applicational field of similarity principle, with which I am concerned. However, the grounds on which it relies deserve discussion. My solution is based on equations developed by Fourier for heat conductibility, whereby temperature and heat are considered to be *sui generis* quantities. It would be, of course, a paradox if a better knowledge of the heat nature — obtained on the basis of molecular theory — would place us in a worse position as to the solution of this particular problem. The solution seems to consist in the fact that Fourier's equations include — for the nature of heat and temperature — an element which ignores Riabouchinsky's argument" [1.32] [15.4]

Speaking of this controversy, Bridgman said that he was "unmoved" by Rayleigh's answer [1.32] [15.5] and that the solution offered by Riabouchinsky is correct; unfortunately, he does not substantiate this statement.

Ehrenfest-Afanassieva demonstrates that the error lies in Riabouchinsky's argument, namely: he failed to make a distinction between quantities of different nature which have the same dimensional equation [1.22] [15.6].

Saint-Guilhem [1.35] carried out dimensional analysis computations on the basis of the system of units with $d^* = 3$.

The dimensional matrix of variables for $d^* = 3$ is:

	l	v	$\Delta\theta$	c	ν	q
L	1	1	2	−3	−1	2
M	0	0	0	1	1	1
T	0	−1	−2	0	−1	−3

(15.23)

The relation, as expressed by physical quantities and by non-dimensional products of powers, is

$$q = kl\Delta\theta\lambda\varphi\left(\frac{lvc}{\lambda}, \quad \frac{\Delta\theta}{v^2}\right). \tag{15.24}$$

Palacios considers however that the problem can be easily solved [1.28]; in his opinion, the correct solution can be found if, based on the variables which describe this physical phenomenon

15.4 p. 127.
15.5 p. 127.
15.6 Ehrenfest-Afanassieva T. A., Dimensional analysis viewed from the standpoint of the theory of similitude, *Phys. Magazine*, 1, 1925.

(15.11) a system of units with $d^* = 4$ fundamental quantities L, M, T, Θ were used. He asks himself, though, whether the list of variables should not be added to the gas constant R – which, of course, is not a problem to be solved by dimensional analysis. Rayleigh neglected the gas constant because, while Palacios's question could be a pertinent one, in the problem under consideration we are faced with a practically imcompressible fluid and hence pressure will not appreciably affect the physical phenomenon [1.32]. Dimensional analysis computations can therefore be resumed on this basis and the dimensional matrix of variables for $d^* = 4$:

	l	v	$\Delta\theta$	c	λ	q
L	1	1	0	−1	1	2
M	0	0	0	1	1	1
T	0	−1	0	−2	−3	−3
Θ	0	0	1	−1	−1	0

(15.25)

The relation, as expressed by the physical quantities and the non-dimensional product of powers, is

$$q = kl\Delta\theta\lambda\varphi\left(\frac{lvc}{\lambda}\right), \tag{15.17}$$

which is the relation derived by Rayleigh.

15.3 Rotation of a Plate within a Fluid Flow

Riabouchinsky maintains that this problem cannot be approached through the analytic method; he therefore suggests a study based on dimensional analysis, considering that this is a typical example of an application.

Let us consider a rectangular plate of sides a and b and turning around one of its symmetry axes; a fluid of density ρ and viscosity η flows at a velocity v in a direction which is normal to the axis, as shown in Fig. 15.1. At low velocities, the plate is maintained in a position normal to the direction of flow ; if, however, the flow velocity exceeds a certain limit value — which is a function of the fluid nature — the plate will be driven in a uniform rotational motion. It is necessary to establish the relation that will yield the angular velocity ω of the rotational motion.

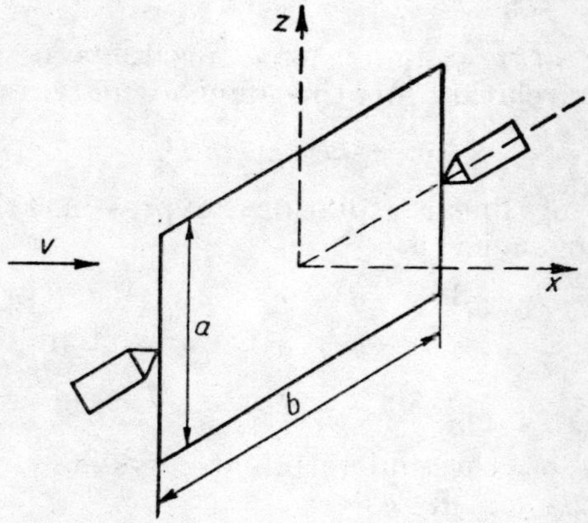

Fig. 15.1 Plate revolution within a fluid flow

Owing to the fact that the plate rotates at a uniform velocity, it could be taken that its moment of inertia I does not affect the physical phenomenon. On the other hand, one should bear in mind the fact that the action of the fluid flow on the plate varies with the various phases of the motion and hence, even though the motion of the plate is periodical it could not be uniform. Thus, when the plate is placed in a normal position with respect to the flow direction, the action of the latter is set off due to symmetry ; the same is true when the plate is placed parallel to the fluid flow ; consequently, accelerations are produced and therefore the plate's moment of inertia I with respect to the axis of rotation should be taken into consideration.

The implicit function of $m^* = 7$ variables is

$$f(l_1, l_2, \rho, \eta, v, \omega, I) = 0. \tag{15.26}$$

The dimensional matrix of variables for $d^* = 3$ is:

$$\begin{array}{c|ccccccc|} & l_1 & l_2 & \rho & \eta & v & \omega & I \\ \hline \mathrm{L} & 1 & 1 & -3 & -1 & 1 & 0 & 2 \\ \mathrm{M} & 0 & 0 & 1 & 1 & 0 & 0 & 1 \\ \mathrm{T} & 0 & 0 & 0 & -1 & -1 & -1 & 0 \end{array} \tag{15.27}$$

whose rank is $r^* = 3$.

The number of non-dimensional products of powers $i^* = 4$. The explicit relation of the dependent variable is

$$\omega = k l_1^a l_2^b \rho^c \eta^d v^e I^g. \tag{15.28}$$

The system of linear equations expressing the condition of dimensional homogeneity is:

$$\begin{aligned} a + b - 3c - d + e + 2g &= 0 \quad &(\mathrm{L})\\ c + d + g &= 0 \quad &(\mathrm{M})\\ -d - e &= -1 \quad &(\mathrm{T}). \end{aligned} \tag{15.29}$$

The solution of the indeterminate system of equations for random unknowns b, d, g is:

$$\begin{aligned} a &= -1 - b - d - 5g \\ c &= -d - g \\ e &= 1 - d \end{aligned} \tag{15.30}$$

while the solution matrix is:

$$\begin{array}{c|ccccccc|} & a & b & c & d & e & & g \\ & l_1 & l_2 & \rho & \eta & v & \omega & I \\ \hline \Pi_1 & 1 & 0 & 0 & 0 & -1 & 1 & 0 \\ \Pi_2 & -1 & 1 & 0 & 0 & 0 & 0 & 0 \\ \Pi_3 & -1 & 0 & -1 & 1 & -1 & 0 & 0 \\ \Pi_4 & -5 & 0 & -1 & 0 & 0 & 0 & 1 \end{array} \tag{15.31}$$

The relation, as expressed by the physical quantities and the non-dimensional products of powers, is

$$\omega = k \frac{v}{l_1} \varphi\left(\frac{l_2}{l_1}, \quad \frac{\eta}{l_1 \rho v}, \quad \frac{I}{l_1^5 \rho}\right). \tag{15.32}$$

The factor of form $\frac{l_2}{l_1}$ shows that relation (15.32) can be used to compare the behaviour of plates of similar sizes. Dimensional analysis, however, cannot allow for the fact that the sizes of the plates' sides will differ from one another. This is why Riabouchinsky substitutes for the non-dimensional product of powers intervening in the inertia moment I, the one given by the ratio of the plate moment of inertia to that of the fluid cylinder generated by the rectangular plate in its rotational motion. Relation (15.32) takes the form:

$$\omega = k \frac{v}{l_1} \varphi\left(\frac{l_2}{l_1},\ Re, \frac{32\, I}{\pi l_1^4 l_2 \rho}\right); \tag{15.33}$$

the non-dimensional product of powers Re, which is the Reynold's number, should include the plate transverse dimension. Relation (15.33) is valid for any rectangular plate. Experimental data have shown that in water or in air, the plate starts rotating when the value of this product of powers is 2; the rotational motion is uniform when $Re = 5$. According to [1.32] [15.7], data obtained through dimensional analysis had to be completed by other results.

Relation (15.33) can also be obtained by dimensional analysis alone; to this end, the Huntley method, based on oriented fundamental quantities is applied. According to this method, let us assume that the axes shown in Fig. 15.1 are fixed and that the phenomenon is symmetrical with respect to plane xz; then, the forces generated by viscosity should be symmetrical with respect to this plane and therefore all that is needed is to take into consideration the components parallel to this plane. According to the law of internal friction stated by Newton, the dynamic viscosity is defined by the relation

$$F = \eta \frac{\mathrm{d}v}{\mathrm{d}n} \mathrm{d}s \tag{15.34}$$

where F = force; η = dynamical viscosity; v = velocity; S = surface area; and n = the distance in the normal direction. From

[15.7] p. 89.

the dimensional equations of physical quantities involved in this relation :

$$\begin{aligned} [F] &= L_{xz}MT^{-2} \\ [v] &= L_{xz}T^{-1} \\ [n] &= L_{xz} \\ [S] &= L_{xz}L_y \end{aligned} \tag{15.35}$$

the dimensional equation of dynamic viscosity results :

$$[\eta] = L_y^{-1}MT^{-1}. \tag{15.36}$$

Now, resuming the dimensional analysis calculations, the dimensional matrix of variables for $d^* = 4$ will be :

(15.37)

	l_1	l_2	ρ	η	v	ω	I
L_{xz}	1	0	−2	0	1	0	2
L_y	0	1	−1	−1	0	0	0
M	0	0	1	1	0	0	1
T	0	0	0	−1	−1	−1	0

The rank of the dimensional matrix $r^* = 4$.

The number of non-dimensional products of powers $i^* = 3$.

The system of linear equations expressing the condition for the dimensional homogeneity is :

$$\begin{aligned} a - 2c + e + 2g &= 0 \quad (L_{xz}) \\ b - c - d &= 0 \quad (L_y) \\ c + d + g &= 0 \quad (M) \\ -d - e &= -1 \quad (T). \end{aligned} \tag{15.38}$$

The solution of the indeterminate system of equations for random unknowns d and g is :

$$\begin{aligned} a &= -1 - d - 4g \qquad & c &= -d - g \\ b &= -g & e &= 1 - d. \end{aligned} \tag{15.39}$$

The solution matrix is :

	a	b	c	d	e		g
	l_1	l_2	ρ	η	v	ω	I
Π_1	1	0	0	0	−1	1	0
Π_2	−1	0	−1	1	−1	0	0
Π_3	−1	−1	−1	0	0	0	1

(15.40)

The relation, as expressed through the physical quantities and the non-dimensional products of powers, is

$$\omega = k \frac{v}{l_1} \varphi\left(\frac{\eta}{l_1 \rho v}, \quad \frac{I}{l_1^4 l_2 \rho} \right); \tag{15.41}$$

it is to be observed that relation (15.33) is of the same form as (15.41), which was derived by dimensional analysis alone.

15.4 Wien's Law

Thermal radiation is that electromagnetic radiation from the inside of a solid walled vacuum chamber that is exclusively thermal and is at thermal equilibrium with the walls. When the solid walls of the chamber are perfectly heat insulated, an even temperature is established after a certain time, as well as a stationary state characterized by the fact that the walls release, in the unit time, electromagnetic energy equal to that absorbed. In such a case the electromagnetic radiation is called "black" radiation and has the same spectral distribution as the radiation released by a body which absorbs all incident radiations.

The fundamental problem of radiation is the determination of the universal function of temperature. For the density of electromagnetic energy, the problem is solved by the law experimentally found by Stefan and mathematically demonstrated by Boltzmann; according to this law, the total density ω of the radiating energy is proportional to the fourth power of absolute temperature θ at which the equilibrium is reached, i.e.

$$\omega + \sigma\theta^4, \tag{15.42}$$

where σ is a dimensional constant. The total density of radiating energy whose frequency ranges from 0 to ∞ is obtained by integrating the radiation spectral density w_ν within the two limits, i.e.

$$w = \int_0^\infty w_\nu \mathrm{d}\nu. \tag{15.43}$$

Quantity w_ν is a function of frequency ν, temperature θ and light velocity in vacuum v. Following Hopf, the relation connecting the physical quantities can be established on the basis of a few simple considerations of dimensional analysis [3.13] [15.8].

From the relation

$$w_\nu d_\nu = \frac{\text{energy}}{\text{surface} \times \text{time}} \tag{15.44}$$

and using the system of units with $d^* = 3$, we derive the dimensional equation for the radiation spectral density :

$$[w_\nu] = \frac{L^2MT^{-2}}{L^2TT^{-1}} = MT^{-2}. \tag{15.45}$$

The implicit function of $m^* = 4$ variables is

$$f(w_\nu, \nu, \theta, v) = 0. \tag{15.46}$$

The dimensional matrix of variables for $d^* = 3$ is :

	w_ν	ν	θ	v
L	0	0	2	1
M	1	0	1	0
T	−2	−1	−2	−1

(15.47)

whose rank is $r^* = 3$.

The number of non-dimensional products of powers $i^* = 1$.

15.8 Hopf, L., Die Naturwissenshaften, VIII, 1920, p. 109.

The explicit relation for the dependent variable is

$$w_\nu = k\,\nu^a\,\theta^b v^c. \tag{15.48}$$

The system of linear equations expressing the condition for the dimensional homogeneity is:

$$\begin{aligned} 2b + c &= 0 \quad &(L)\\ b &= 1 \quad &(M)\\ -a - 2b - c &= -2 \quad &(T). \end{aligned} \tag{15.49}$$

The solution for the determinate system of equations is:

$$a = 2, \qquad b = 1, \qquad c = -2 \tag{15.50}$$

and the solution matrix is:

$$\begin{array}{c|cccc|} & & a & b & c\\ & w_\nu & \nu & \theta & v\\ \hline \Pi & 1 & -2 & -1 & 2 \end{array} \tag{15.51}$$

The relation, as expressed by the physical quantities, is

$$w_\nu = k\,\frac{\nu^2\theta}{v^2}. \tag{15.52}$$

From relations (15.42), (15.43) and (15.52), we can write

$$\sigma\theta^4 = k\int_0^\infty \frac{\nu^2\theta}{v^2}\,d\nu, \tag{15.53}$$

from which it follows that the numerical factor of proportionality k depends on variables σ, θ, ν and v. We again have recourse to dimensional analysis to derive the relation connecting these variables. The dimensional equation of variable σ is obtained on the basis of Stefan-Boltzmann's law. It follows from (15.42) and (15.43) that

$$\sigma = \frac{\int_0^\infty w_\nu \mathrm{d}\nu}{\theta^4}; \tag{15.54}$$

consequently, the dimensional equation

$$[\sigma] = \frac{\dfrac{\text{energy}}{\text{surface} \times \text{time}}}{\text{energy}^4} = \frac{1}{\mathrm{L^2T(L^2MT^{-2})^3}} = \mathrm{L^{-8}M^{-3}T^5}. \tag{15.55}$$

The implicit function of $m^* = 5$ variables is

$$f(k,\ \sigma,\ \theta,\ \nu,\ v) = 0. \tag{15.56}$$

The dimensional matrix of variables for $d^* = 3$ is:

	k	σ	θ	ν	v
L	0	−8	2	0	1
M	0	−3	1	0	0
T	0	5	−2	−1	−1

(15.57)

whose rank is $r^* = 3$.

The number of non-dimensional products of powers $i^* = 1$. The explicit relation for constant k [3.13] is

$$k = f\sigma^a \theta^b \nu^c v^d. \tag{15.58}$$

The system of linear equations expressing the condition for dimensional homogeneity is:

$$\begin{aligned} -8a + 2b + d &= 0 \quad (\mathrm{L}) \\ -\,3a + b &= 0 \quad (\mathrm{M}) \\ 5a - 2b - c - d &= 0 \quad (\mathrm{T}). \end{aligned} \tag{15.59}$$

The solution of the indeterminate system of equations for the random unknown a is:

$$b = 3a, \quad c = -3a, \quad d = 2a. \tag{15.60}$$

The solution matrix is:

$$\begin{array}{c|cccc|} & a & b & c & d \\ & \sigma & \theta & \nu & v \\ \hline \Pi & 1 & 3 & -3 & 2 \end{array} \tag{15.61}$$

The relation, as expressed by the dimensional product of powers, is

$$k = \varphi\left(\sigma \frac{\theta^3 v^2}{\nu^3}\right) = f'\left(\frac{\nu}{\theta}\right). \tag{15.62}$$

Relations (15.52) and (15.62) lead to relation

$$w_\nu = \frac{\nu^2 \theta}{v^2} f'\left(\frac{\nu}{\theta}\right) \tag{15.63}$$

or

$$w_\nu = \frac{\nu^2}{v^2}\, \theta \, \frac{\nu}{\theta} f\left(\frac{\nu}{\theta}\right) \tag{15.64}$$

wherefrom we derive immediately

$$w_\nu = \frac{\nu^3}{v^2} f\left(\frac{\nu}{\theta}\right) \tag{15.65}$$

which is the expression for Wien's law.

15.5 The Way in which a Sphere Falls through a Viscous Medium

Bridgman made use of dimensional analysis to study the problem enunciated by Stokes with reference to the fall, under its own weight, of a sphere through a viscous medium; the sphere is small compared with the fluid volume. The physical model of the phenomenon has already been described (Section III.13.1). According to Bridgman, the non-dimensional product of powers given by the ratio of the fluid to the sphere densities is:

$$\Pi_1 = \frac{\rho}{\rho_0}. \tag{15.66}$$

From the variables considered in the calculations of dimensional analysis, he eliminates the fluid density. Then, the implicit function of $m^* = 5$ variables is:

$$f(d,\ g,\ \rho_0,\ \eta,\ v) = 0. \tag{15.67}$$

The dimensional matrix of variables for $d^* = 4$ is:

	d	g	ρ_0	η	v
L	1	0	−3	−2	1
M	0	−1	1	0	0
T	0	0	0	1	−1
F	0	1	0	1	0

(15.68)

whose rank is $r^* = 4$.

The number of non-dimensional product of powers $i^* = 1$.

The explicit relation for the dependent variable is:

$$v = kd^a g^b \rho_0^c \eta^e. \tag{15.69}$$

The system of linear equations expressing the condition of dimensional homogeneity is:

$$\begin{aligned} a - 3c - 2e &= 1 \quad (\text{L}) \\ -b + c &= 0 \quad (\text{M}) \\ e &= -1 \quad (\text{T}) \\ b + e &= 0 \quad (\text{F}). \end{aligned} \tag{15.70}$$

And the solution of the determinate system of equations :

$$a = 2, \quad b = 1, \quad c = 1, \quad e = -1. \tag{15.71}$$

The solution matrix is :

	a	b	c	e	
	d	g	ρ_0	η	v
Π	-2	-1	-1	1	1

(15.72)

The relation, as expressed by the physical quantities and the non-dimensional product of powers given before is :

$$v = k \frac{d^2 g \rho_0}{\eta} \varphi\left(\frac{\rho}{\rho_0}\right). \tag{15.73}$$

Bridgman is also of the opinion that for dimensional analysis computations the function with $m^* = 6$ variables can be used, together with the system of units with $d^* = 3$ fundamental quantities.

The implicit function of $m^* = 6$ variables is

$$f(d,\ g,\ \rho_0,\ \rho,\ \eta,\ v) = 0. \tag{15.74}$$

The dimensional matrix for $d^* = 3$ is :

	d	g	ρ_0	ρ	η	v
L	1	1	-3	-3	-1	1
M	0	0	1	1	1	0
T	0	-2	0	0	-1	-1

(15.75)

whose rank is $r^* = 3$.

The number of non-dimensional products of powers $i^* = 3$.

The relation, as expressed by the physical quantities and the non-dimensional products of powers, is

$$v = k \frac{\eta}{d \rho_0} \varphi\left(\frac{\rho}{\rho_0},\ \frac{d^3 g \rho_0^2}{\eta}\right). \tag{15.76}$$

Palacios considers that this physical phenomenon involves the apparent weight of the sphere

$$G = \frac{1}{6} \pi D^2(\rho_0 - \rho)g \tag{13.1}$$

and, as a consequence, this should be included in the list of variables.
The implicit function of $m^* = 5$ variables is

$$f(d, G, \rho, \eta, v) = 0. \tag{15.77}$$

The dimensional matrix for $d^* = 3$ is :

	d	G	ρ	η	v
L	1	1	−3	−1	1
M	0	1	1	1	0
T	0	−2	0	−1	−1

(15.78)

whose rank s $r^* = 3$.
The number of non-dimensional products of powers $i^* = 2$.
The explicit relation of the relevant variable is

$$G = Kd^a \rho^b \eta^c v^e. \tag{15.79}$$

The system of linear equations expressing the condition of dimensional homogeneity is :

$$\begin{aligned} a - 3b - c + d &= 1 \quad \text{(L)} \\ b + c &= 1 \quad \text{(M)} \\ - c - d &= -2 \quad \text{(T).} \end{aligned} \tag{15.80}$$

The solution of the indeterminate system of equations for the random unknown b is :

$$a = 1 + b, \quad c = 1 - b, \quad d = 1 + b. \tag{15.81}$$

And the solution matrix is :

$$\begin{array}{c|ccccc|}
 & a & b & & c & e \\
 & d & G & \rho & \eta & v \\
\hline
\Pi_1 & -1 & 1 & 0 & -1 & -1 \\
\Pi_2 & 1 & 0 & 1 & -1 & 1
\end{array} \qquad (15.82)$$

The relation, as expressed by the physical quantities and the non-dimensional product of powers is :

$$G = kd\eta v \varphi \left(\frac{d\rho v}{\eta} \right) \qquad (15.83)$$

or else, put under another form to make explicit the independent variable :

$$v = k \frac{G}{d\eta} \varphi \left(\frac{d\rho v}{\eta} \right), \qquad (15.84)$$

the product of powers within the brackets being the Reynolds number. Relations (13.1) and (15.84) lead to the final relation

$$v = k \frac{d^2(\rho_0 - \rho)g}{\eta} \varphi \left(\frac{d\rho v}{\eta} \right) \qquad (15.85)$$

which is in good agreement with the relation derived through the analytic method.

16 Determination of the Factors and of the Exponents Values

16.1 Using a Graph

Dimensional analysis computations carried out for a given function of variables could lead to :

— a relation between physical quantities, when $i^* = 1$;

— a relation between non-dimensional products of powers, when $i^* > 1$.

These computations however cannot supply either the values of the numerical factor of proportionality, or the values of non-dimensional products of powers. The results are sought, instead, either from experiments or by having recourse to the theory of physical phenomena.

To illustrate, let us consider a swirl-chamber injector with a frontal return loop. The fuel enters the swirlchamber whose diameter is D, to be subsequently pressed through a nozzle of a diameter d. The fuel has a density ρ, a dynamic viscosity η and a superficial tension σ. The total rate of fuel flow through the swirlchamber is B_t, while that flowing through the nozzle is B_r; their ratio

$$n = \frac{B_t}{B_r} \tag{16.1}$$

is called the recycling ratio. The equivalent velocity v at which the fuel flows through the nozzle is given by the obvious equation

$$v = \frac{4B_r}{d^2}. \tag{16.2}$$

It is necessary to find the relation for the coefficient of the real rate of flow ξ. As can be inferred, this physical phenomenon is affected by the recycling ratio n and the list of variables should also include the coefficient of the ideal rate of flow ξ_0. The independent variable will be the ratio between the rate of flow coefficients. As a consequence, in (3.7) this relation is written

$$\frac{\xi}{\xi_0} = f(D,\ d,\ \rho,\ \eta,\ \sigma,\ n,\ v), \tag{16.3}$$

on which will rely the dimensional analysis computations.

The relation, as expressed by the non-dimensional products of powers, is

$$\frac{\xi}{\xi_0} = \varphi\left(\frac{D}{d},\ \frac{d\rho\sigma}{\eta^2},\ \frac{d\rho v}{\eta},\ n\right). \tag{16.4}$$

Since relation (16.3) was taken into consideration, in relation (16.4) the recycling ratio n was introduced. The last non-dimensional product of powers is the Reynolds number.

Such a typical injector was tested with water and with a water-glycerine mixture. The experimental determinations made when the injector was operated with the return loop closed, i.e. $n = 1$, for liquids having the kinematic viscosity $\nu = 1\ldots2.5°E$, are represented in Fig. 16.1.

For the assumption made, relation (16.4) takes the form:

$$\frac{\xi}{\xi_0} = k\left(\frac{D}{a}\right)^a \Pi^e (Re)^g \tag{16.5}$$

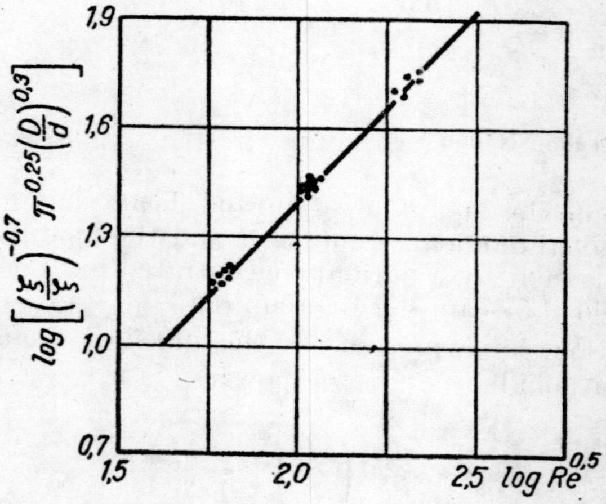

Fig. 16.1 Experimental points during injector operation

where

$$\Pi = \frac{d\rho\sigma}{\eta^2}. \tag{16.6}$$

It will be seen that the points on the diagram follows a unitary-slope straight line; the linear dependence was obtained by checking the various values for the non-dimensional products of powers a, e

and g. As a consequence, the relation established by means of dimensional analysis and using the data from Fig. 16.1 will take the form

$$\frac{\xi}{\xi_0} = 11.21 \left(\frac{D}{a}\right)^{0.429} \Pi^{0.357} (Re)^{-0.76} . \tag{16.7}$$

As pointed out by some authors [3.7], relation (16.7) holds true for:

$$\begin{gathered} 3{,}880 < Re < 48{,}100 \\ 380 < \Pi < 217{,}000 \\ 5.33 < \frac{D}{d} < 8. \end{gathered} \tag{16.8}$$

16.2 The Algebraic Method

Let us consider again a supported beam having a length l, the cross-sectional moment of inertia I and the modulus of elasticity E; the beam is bent by a normal concentrated force P. The relation which will yield the beam sag s is required.

The relation, as expressed by the physical quantities and by the non-dimensional products of powers, is

$$s = kl \left(\frac{I}{l^4}\right)^b \left(\frac{l^2 E}{P}\right)^c . \tag{16.9}$$

Relation (16.9), reached through dimensional analysis, includes three unknowns: the numerical factor of proportionality k and the exponents of the non-dimensional products of powers b and c. To find the values of these unknowns, data supplied – theoretically – by three experimental measurements are used; these are given in Table 16.1. On the basis of the data, supplied in this table, the values for the non-dimensional products of powers given in relation (16.9) are estimated; the results of this estimation are given in Table 16.2.

TABLE 16.1. **Measured values for beam bending**

No. of order	l	I	E	P	s
1	500	500	2.1×10^6	3,700	9.2
2	900	900	2.1×10^6	2,100	16.9
3	1,700	4,500	2.1×10^6	900	9.7

The values of quantities l and s, as given in Table 16.1, are introduced in relation (16.9) along with the non-dimensional products of powers taken from Table 16.2; the result will be the following system of three equations with three unknowns:

TABLE 16.2

Values of non-dimensional products of powers

No. of order	$\frac{I}{l^4}$	$\frac{l^2E}{P}$
1	8.00×10^{-9}	0.14×10^9
2	12.70×10^{-9}	0.81×10^9
3	0.54×10^{-9}	6.75×10^9

$$\begin{aligned} 9.2 &= 500\quad k(8.00\times10^{-9})^b\,(0.14\times10^9)^c \\ 16.9 &= 900\quad k(12.70\times10^{-9})^b\,(0.81\times10^9)^c \\ 9.7 &= 1{,}700\quad k(0.54\times10^{-9})^b\,(6.75\times10^9)^c. \end{aligned} \tag{16.10}$$

The determinate system of equations admits the solution:

$$k \approx 0.02, \quad b \approx -1, \quad c \approx -1. \tag{16.11}$$

Substituting the above values in relation (16.9) results in

$$s = 0.02\,\frac{l^3P}{IE} \tag{16.12}$$

which is practically identical with the relation obtained by the analytic method.

★

In order to use the theory of the phenomenon, according to [1.32] [16.1] the relation (16.9) for the sag of a supported beam under the action of a central, concentrated, vertical force is considered; consequently, data supplied by dimensional analysis in this case are rather indefinite.

As is known, from Hooke's law, the elastic strain in a body and the force which caused it are connected by a proportionality relationship: hence relation (16.9) will become

$$s = k \frac{P}{lE} \left(\frac{I}{l^4} \right)^b. \tag{16.13}$$

It follows from the theory of elasticity, that the above phenomenon is also affected by the moment of inertia of the cross-section and that the strain is proportional to the length; hence relation (16.13) takes the form

$$s = k \frac{l^3 P}{IE}. \tag{16.14}$$

17 Checking Physical Relationships

17.1 Correct Relationships

Any mathematical calculus whose final phase is a relation between the variables used to describe a certain physical phenomenon must be correct as far as its dimensional homogeneity is concerned. Since the non-observance of this requirement will lead to errors, one should always make sure that the dimensional homogeneity is satisfied. This will not involve going through all phases of dimensional analysis calculations as is usual when deriving the relations between physical quantities and non-dimensional products of powers. All that is needed is merely to introduce — in the relation under con-

---

[16.1] p. 74.

sideration — the dimensional equations of variables, written in a coherent system of units. To check dimensional homogeneity is, of course, an easy task; the importance of such checks can never be overemphasized, whether they are carried out along the way or, and above all, on the final result. Some specialists consider that the checks conducted on dimensional homogeneity are similar to the "proof of nine" often carried out to assess the correctness of arithmetic calculation. As already shown in (Section I.6.1) one of the many applications of dimensional equations is precisely this check.

In about 1775, Chezy suggested the first relation with a potential applicability to the loss of unit load J as a consequence of the uniform motion of currents with the free surface; this was of the form

$$J = k\,\frac{v^2}{C^2 R}, \tag{17.1}$$

where v is the current velocity, C is the bed roughness while R is the hydraulic radius. The dimensional equation of variable C is

$$[C] = \mathrm{L}^{1/2}\mathrm{T}^{-1}. \tag{17.2}$$

In 1865, Darcy suggested a relation to express the loss of unit load through a circular duct with a constant cross-section; this was of the form

$$J = k\,\frac{v^2}{gd}, \tag{17.3}$$

where the symbols represent the flow velocity v, the normal acceleration in free fall g and the duct diameter d [2.18] [17.1].

It can easily be seen that these relations — which verify all experimental data — are also correct from the standpoint of dimensional homogeneity.

Let us consider, for example, Bernoulli's relation:

$$\frac{p}{\gamma} + \frac{v^2}{2g} + z = K \tag{17.4}$$

where p is the pressure; v is the velocity; z is the height; γ is the specific weight; and g is the normal acceleration in free fall.

[17.1] p. 303.

The dimensional equations of variables result from the following dimensional matrix for $d^* = 3$:

$$\begin{array}{c|ccccc|} & p & v & z & \gamma & g \\ \hline \mathrm{L} & -1 & 1 & 1 & -2 & 1 \\ \mathrm{M} & 1 & 0 & 0 & 1 & 0 \\ \mathrm{T} & -2 & -1 & 0 & -2 & -2 \end{array} \qquad (17.5)$$

As can be seen, each left-hand term of relation (17.5) has the dimensional equation of length; then, obviously, the constant K in the right-hand side of the equation should have the same dimensional equation.

17.2 Erroneous Relations

Owing to the fact that dimensional homogeneity was sometimes neglected, erroneous relations were derived. Thus, for instance, relation [17.2]

$$\log(C_1 - C_0) = \frac{D}{d}\, t \qquad (17.6)$$

was suggested, where C_1 and C_0 are two concentrations of the same solution; D is the diffusion; t is the diffusion time; and d is the thickness of a layer having a certain concentration [2.24].

Obviously, relation (17.6) is afflicted by two errors: first, the left-hand member is a logarithm and hence the difference will be a number which should not have a dimensional equation; second, the right-hand member has the dimensional equation for length and it should not have a dimensional equation. By using the dimensional analysis, it can be easily shown that relation (17.6) should be written as

$$\log \frac{C_1 - C_0}{C_0} = \frac{D}{d^2}\, t \qquad (17.7)$$

which is the correct form.

[17.2] *Zeitschrift für Physik, Chemie*, XXXXV, 1903, p. 218.

IV. General Dimensional Analysis

18 The Basis of General Dimensional Analysis

18.1 Basic Problem of General Dimensional Analysis

The study of any physical phenomena by restricted dimensional analysis includes three stages, namely :

— description of the physical phenomenon and specification of physical quantities associated with the development of that phenomenon ;

— dimensional analysis computations which can lead to a relation between physical quantities when $i^* = 1$, or to a relation between non-dimensional products of powers, when $i^* > 1$;

— determination of the values of the numerical factor and of the exponents of non-dimensional products of powers.

Information yield from dimensional analysis computations becomes more vague when the number of non-dimensional products of powers increases. Maximum efficiency of dimensional analysis is reached for the unit value of non-dimensional products of powers. In any other situations, this knowledge requires additional information obtained from experimental data or derived from the phenomenon theory.

Let us consider the following question : is it possible to obtain additional information by dimensional analysis computations when $i^* > 1$? It goes without saying that "additional information" refers to the values of the exponents of non-dimensional products of powers, since dimensional analysis cannot determine the value of a numerical factor of proportionality from a physical relation. This question is the basic problem of general dimensional analysis. It is the author's belief that the question may be answered affirmatively and this, indeed, is the purpose of this chapter.

18.2 Remarks on the Π Theorem and on Rayleigh's Method

Remark 1 refers to the Π theorem. The function of physical quantities involved in any definite physical phenomenon is considered to be:

$$f(A,\ B,\ C,\ \ldots,\ M,\ N,\ P,\ \ldots) = 0. \tag{3.2}$$

The dimensional equations of the physical quantities A, B, C, ... M, N, written according to a system of units of measurement with d^* fundamental quantities (L, M, T) result from the following dimensional matrix:

$$\begin{array}{c|cccc|} & A & B & C\ldots & P\ldots \\ \hline \mathrm{L} & l_a & l_b & l_c & l_p \\ \mathrm{M} & m_a & m_b & m_c & m_p \\ \mathrm{T} & t_a & t_b & t_c & t_p \end{array} \tag{9.1}$$

From the Π theorem, non-dimensional products of powers are written on the basis of variables A, B, C, ... chosen at random:

$$\Pi_m = \frac{M}{A^{a_m}B^{b_m}C^{c_m}\ \ldots},\quad \Pi_n = \frac{N}{A^{a_n}B^{b_n}C^{c_n}\ldots}\ \ldots, \tag{8.2}$$

The dimensional homogeneity condition of a non-dimensional product, for example Π_p, is expressed by the system of linear algebraic equations:

$$\begin{aligned} l_a \cdot a_p + l_b \cdot b_p + l_c \cdot c_p + \ldots &= l_p \\ m_a \cdot a_p + m_b \cdot b_p + m_c \cdot c_p + \ldots &= m_p \\ t_a \cdot a_p + t_b \cdot b_p + t_c \cdot c_p + \ldots &= t_p. \end{aligned} \tag{18.1}$$

Taking into account relations (8.2) where the values of unknowns, as obtained from the systems of equations (9.2), are replaced, relation (3.2) becomes

$$\varphi(\Pi_m, \Pi_n, \Pi_p, \ldots) = 0 \qquad (8.1)$$

where a smaller number of variables Π_m, Π_p ... is involved.

Remark 1 underlines the fact that the *sine qua non* condition prevailing whenever a non-dimensional product of powers is established requires that the system of equations between the dimension exponents and the quantity exponents, which express the dimensional homogeneity condition of the product, is a determinate one. Any of the applications already discussed in this work (Section III) confirms this situation. The condition is needed if it is desired to have for the system of non-homogeneous equations (18.1) with d^* equations the same number of unknowns, that is to have a Cramer system. It is known that, any system of Cramer equations has but one solution derived from Cramer's formulae :

$$a = \frac{\det X_a}{\det X}, \qquad b = \frac{\det X_b}{\det X}, \qquad c = \frac{\det X_c}{\det X} \qquad (18.2)$$

where det $X_a, \ldots$ (det being short for determinant) are the matrices derived from the matrices of coefficients of unknowns X by replacing the coefficients of unknowns l_a, m_a, $t_a, \ldots$ by the free term l_p, m_p, t_p [4.1].

Remark 2 also refers to the Π theorem. Let us consider, for example, the physical phenomenon of the pressure drop in a circular pipe having a constant cross-section; this phenomenon may be described, according to Darcy's relation, by the following function of m^* physical quantities, where $m^* = 6$

$$f(\Delta p, \gamma, v, l, g, d) = 0. \qquad (18.3)$$

If variables Δp, v and d are admitted as fundamental, three non-dimensional products of powers might be written, namely :

$$\Pi_\gamma = \frac{\gamma}{\Delta p^{a_\gamma} v^{b_\gamma} d^{c_\gamma}}, \qquad \Pi_l = \frac{l}{\Delta p^{a_l} v^{b_l} d^{c_l}}, \qquad \Pi_g = \frac{g}{\Delta p^{a_g} v^{b_g} d^{c_g}} \qquad (18.4)$$

Computing the first non-dimensional product of powers for the $d^* = 3$ assumption results in

$$\Pi_\gamma = \frac{\Delta p}{\gamma d}. \tag{18.5}$$

Relation (18.5) includes the product of variables γ and d; these variables are also included in the correct relation —which is known — through their quotient

$$\Delta p = k\,\frac{\gamma v^2 l}{g d}. \tag{18.6}$$

Remark 2 underlines the fact that the choice of variables considered fundamental for the case under study is arbitrary. On the basis of this choice non-dimensional products of powers corresponding to the other variables are written. No additional condition is required in deriving these variables, which means that only their dimensional correctness is taken into consideration. In such a situation, it is obvious that some non-dimensional products of powers might not shown the correct physical relation between the variables.

Remark 3 refers to the Rayleigh methods, according to which the relation between the variables in function (3.2) is written in the form

$$P = kA^a B^b C^c \ldots M^m N^n \ldots \tag{8.17}$$

that is only one non-dimensional product of powers is written. The system of algebraic equations

$$\begin{aligned} l_a \cdot a + l_b \cdot b + l_c \cdot c + \ldots l_m \cdot m + l_n \cdot n + \ldots &= l_p \\ m_a \cdot a + m_b \cdot b + m_c \cdot c + \ldots m_m \cdot m + m_n \cdot n + \ldots &= m_p \\ t_a \cdot a + t_b \cdot b + t_c \cdot c + \ldots t_m \cdot m + t_n \cdot n + \ldots &= t_p \end{aligned} \tag{9.2}$$

expressing the condition of dimensional homogeneity may be either a determined one (when $n^* = r^*$), or a non-determined one (when $n^* > r^*$). In the first case, the system of equations is of the Cramer type and hence it admits only one solution; in the second case, the system of equations is a non-determined one and admits a general solution which is expressed as a function of some arbitrary unknowns. Dimensional analysis computation leads either to a relation involving physical quantities or to one involving non-dimensional products of powers.

In the case considered above, this relation will be written in the form

$$\Delta p = k\gamma^a v^b l^c g^e d^h \tag{18.7}$$

and dimensional analysis computations lead to the relation:

$$\Delta p = k\gamma l \left(\frac{lg}{v^2}\right)^e \left(\frac{d}{l}\right)^h; \tag{18.8}$$

three non-dimensional products of powers are involved in the above relation, which obviously means that the information supplied is vague.

Remark 3 therefore underlines the fact that with Rayleigh's method, while writing initially only one non-dimensional product of powers (8.17), when $n^* > r^*$ the dimensional analysis computations lead to a relation involving several non-dimensional products of powers.

For the solution of the indeterminate system of equations (9.2), the arbitrary choice of unknowns considered to be independent may lead to non-dimensional products of powers incorrect with respect to their physical interrelations. Thus, for instance, the correct relation (18.6) contains the quotient of variables l and g, while the non-dimensional product of powers from relation (18.8) contains their product. In this respect it can be seen that the result is similar to that which led to the second remark referring to the Π theorem.

This remark might be completed by the statement that for a given function of m^* physical quantities, the Π theorem could lead to a smaller or greater i^* number, as determined by number d^*, which is chosen arbitrarily. Of course, if $d^* - s$ with adequate values were available, the result would always be $i^* = 1$, which is the most complete result obtained through the dimensional method of study. It is known, however, that the numbers d^* presently used have optimized values and therefore no increase in these values is recommended.

18.3 Implicated Relations

Let us consider the relation between variables involved in the development of a given physical phenomenon, written according to Rayleigh's method

$$P = kA^aB^bC^c \ldots M^mN^n \ldots \tag{8.17}$$

The explicit variable P is generally the independent variable; in some cases, any other variable can be put in an explicit form, depending on the specific goal. The form of this relation is the usual one; it is the simplest form for the corresponding explicit quantity. However, it is obvious that relation (8.17) can be written in another form as well. Thus, for instance, relation (8.17) may be raised to a higher power without losing its physical validity; in this way, other relations can also be derived, all implicated by the same relation as expressed in its usual form. The implicated relations differ from one another only in form. Raising a physical relation to any power might be of interest, for instance in the case when the values taken by the quantity exponents are rational numbers; in practice, a relation is raised to such a power as to obtain integral values for its exponents. To obtain such values, the relation considered should be raised to the power of a common multiple of the denominators of all rational quantity exponents. When the common multiple chosen is the least common multiple, the new values of the quantity exponents will be minimum.

Assuming as given a relation of the form (8.17) — for instance Toricelli's relation $v = (2gh)^{\frac{1}{2}}$ whereby the *in vacuo* velocity of fall of a body v is given as a function of the normal acceleration of free fall g and of the height of fall h — the values of the quantity exponents will therefore be known too. When raised, for instance, at the second power, this relation takes the form

$$v^2 = 2gh, \tag{18.9}$$

where the quantity exponents take new values, the least integral values. However, there is no reason whatsoever why we should not write relation (3.13) in, say, the form

$$v^8 = 16g^4h^4, \tag{18.10}$$

where quantity exponents, while still integers, are not minimum. Any relation of the type (18.10) is more complex and of no practical interest.

Assuming that the form of relation (8.17) is not known and it is sought to derive it through dimensional analysis, the values of quantity exponents a, b, c, ... will represent the solution of a system of equations between the dimensional and the quantity exponents, expressing the dimensional homogeneity conditions (9.2). As given, the singular values of the quantity exponents can be obtained only when the system of equations (9.2) is a determinate one; if indeterminate, a general solution is accepted which leads to an infinity of solutions, among which the solution of the correct physical relation is to be found. The problem studied by dimensional analysis is to select this singular solution. The infinity of solutions which satisfy a given indeterminate system of equations (9.2) also includes the solutions whose values of the quantity exponents hold true for all the relations implicated by the relation written in the standard form.

Making a variable explicit — as attempted by Rayleigh's method — means assigning to the corresponding quantity exponent a unit value; this leads to a determinism or, in other words, to the individualization of a certain solution which should then be selected from an infinity of solutions. The selection is, of course, a difficult task since the solutions might be integral and rational, positive and negative.

It should be noted that implicated relations differing only in form can be written and that the effect of putting a quantity in an explicit form is to individualize the values of the remaining quantity exponents — **Remark 4**

18.4 Equations Expressing the Condition of Dimensional Homogeneity

Because of Remarks 1 and 3, it is considered that an investigation into the systems of equations expressing the dimensional homogeneity condition of physical relations is necessary; the object of this investigation is twofold : to avoid the condition required for the application of the Π theorem; and to eliminate the situation which leads to several non-dimensional products of powers when applying Rayleigh's method.

According to the second part of the Π theorem, the number of non-dimensional products of powers is given by

$$i^* = m^* - r^* \qquad (8.8)$$

where m^* is the number of physical quantities which describe a determined physical phenomenon and r^* is the rank of the dimensional matrix. The value of number r^* is correlated with the number of fundamental quantities d^*, that is

$$r^* = f(d^*). \tag{18.11}$$

In the case of indeterminate systems of equations which express the dimensional homogeneity condition, $i^* - 1$ exponents of non-dimensional products of powers remain unknown; their number will therefore be given by the relation

$$i^* - 1 = m^* - r^* - 1. \tag{18.12}$$

As given, general dimensional analysis is used in those cases when restricted dimensional analysis computation cannot ascribe singular values to the quantity exponents. Indeed, in the case of a given function of m^* physical quantities, it would be necessary to raise the rank of the dimensional matrix; but this increase is neither possible for a given system of units of measurement nor always satisfactory when changing the system of the units. As a consequence, the only possible way of finding an answer to the problem of general dimensional analysis is to adequately solve the system of equations expressing the dimensional homogeneity condition. With this end in view, let us consider the systems of equations relating the dimensional to the quantity exponents.

The equations expressing the dimensional homogeneity condition are written for each fundamental quantity. As given, the number d^* of these equations is correlated to the system of units adopted and can be at most equal to the number of the fundamental quantities d^*; which is more often than not. Owing to the fact that in some fields of science a variety of systems of units with different numbers of fundamental quantities are available, the number of (9.2) equations to be obtained for a given function (3.2) can vary too. In mechanics, for instance, three systems of units of measurement are available in practice, namely with : $d^* = 2$ fundamental quantities, L, T ; $d^* = 3$ fundamental quantities, L, M, T ; and $d^* = 4$ fundamental quantities, L, M, T, F.

The coefficients from equations (9.2) represent the dimensional exponents of the dimensional equations derived for the physical quantities. For instance, the dimensional exponents from the dimensional equations established for quantity A, pertaining to the field of mechanics :

$$[A] = \mathrm{L}^{l_a}\mathrm{M}^{m_a}\mathrm{T}^{t_a} \tag{5.45}$$

are l_a, m_a, t_a. As shown already (Section I.1), physical quantities may be classified into primary and secondary ones, according to their physical significance. The former are immediate, intuitive concepts, with no relationships between them. The latter are defined through a variety of relations. Thus, for instance, the area of a rectangle S is defined by the relation

$$S = l_1 \cdot l_2, \tag{2.7}$$

where l_1 and l_2 are the sides of the rectangle. The density of an homogeneous body is defined by relation

$$\rho = \frac{m}{V}, \tag{18.13}$$

where m is the mass of the body, and V its volume. For the general case, the velocity is defined by the relation

$$v = \frac{\mathrm{d}s}{\mathrm{d}t}, \tag{18.14}$$

where ds is the space travelled and dt, the time taken to travel that space. The force impulse is expressed by the relation :

$$I = \int F \mathrm{d}t, \tag{18.15}$$

where F is the force and dt is the force active time. Secondary quantities are defined by means of either primary or other secondary quantities, previously defined (Section I. 1.2); relations used to define secondary quantities include such algebraic operations as multiplication and division, differentiation and integration. As shown before (Section I.2), physical quantities are also classified as fundamental and derived. Fundamental quantities are chosen at random; derived quantities are defined by dimensional equations written as a function of the fundamental quantities. Physical quantities are now described by dimensional equations including dimensional exponents of various values. When defining secondary quantities, derivatives of a non-integer order are not used. For illustrative purposes, several physical quantities used in mechanics are given in Table 18.1, along with their dimensional equations written in the three systems of units used in this field.

TABLE 18.1. **Dimensional equations of some physical quantities in mechanics**

No. of order	Physical quantity	System of units of measurement		
		LT	LMT	LMTF
1	Velocity	LT^{-1}	LT^{-1}	LT^{-1}
2	Acceleration	LT^{-2}	LT^{-2}	LT^{-2}
3	Mass	L^3T^{-2}	M	M
4	Force	L^4T^{-4}	LMT^{-2}	F
5	Gravitational constant	—	$L^3M^{-1}T^{-2}$	$L^2M^{-2}F$
6	Dynamic constant	—	—	$L^{-1}T^2M^{-1}F$
7	Moment of force	L^5T^{-4}	L^2MT^{-2}	LF
8	Mechanical work	L^5T^{-4}	L^2MT^{-2}	L^2MT^{-2}
9	Energy	L^5T^{-4}	L^2MT^{-2}	L^2MT^{-2}
10	Power	L^5T^{-5}	L^2MT^{-3}	$LT^{-1}F$
11	Impulse	L^4T^{-3}	LMT^{-1}	LMT^{-1}
12	Percussion	L^4T^{-3}	LMT^{-1}	TF
13	Moment of inertia	L^5T^{-2}	L^2M	L^2M
14	Density	T^{-2}	$L^{-3}M$	$L^{-3}M$

As can be seen, dimensional exponents when referred to fundamental quantities take significant values which can be either integers or zero, positive or negative, contained within a restricted range.

In other fields of science, dimensional exponents take irreducible integral and rational values, positive and negative; the numerator and the denominator of rational dimensional exponents have values contained within the same limits.

The unknowns included in the equations expressing the dimensional homogeneity condition represent the quantity exponents corresponding to the variables used to describe the physical phenomenon. The number n^x of unknowns is given by the number of variables and is smaller than the latter by one single unit because the quantity made explicit by a relation written according to Rayleigh's method has its quantity exponent equal to unity. Since in some fields of science systems of units having different numbers of fundamental quantities are available, values e^* and n^* are less representative than the characteristic difference $n^* - e^*$. Between the number of equations and the number of unknowns the following relation exists:

$$e^* \gtrless n^*. \tag{18.16}$$

The values of the quantity exponents are computed from the values of the dimensional exponents, by applying elementary algebraic operations. Hence, their values will be of the like nature, that is the quantity exponents can take integral and rational, positive and negative values.

The terms of equations (9.2) represent products of a dimensional and a quantity exponent, i.e. products of a numerical value and an unknown. The number of terms t^* of any equation is given by the number of physical quantities used to describe the relevant physical phenomenon, the dimensional equations of which include the fundamental quantity corresponding to the equation considered. To describe a physical phenomenon of, say, mechanics, use is made of some quantities pertaining both to the fields already constituted, such as geometry and kinematics, and to the field of mechanics. It is assumed that the dimensional equations of the m^* physical quantities are written in the system of units with $d^* = 3$ fundamental quantities: L, M, T. If the dimensional equations of all physical quantities applied in geometry, kinematics and mechanics are considered, it will be found that the fundamental quantity L is included in the dimensional equations written for all geometric quantities, as well as for the great majority of the kinematic and mechanical ones; the fundamental quantity T is included in the dimensional equations written for all kinematic quantities and for some of the mechanical ones, while the fundamental quantity M is included only in the dimensional equations written for the mechanical quantities. For any given function of m^* variables, let m^* be the number of quantities the dimensional equations of which include a certain fundamental quantity. Hence, for the system of units with $d^* = 3$ fundamental quantities for instance, there will be three numbers, m^*_{L}, m^*_{M} and m^*_{T}, representing the number of quantities the dimensional equations of which contain the fundamental quantities, i.e. L, T and M respectively. Each of the three numbers obey relation

$$m^*_{\mathrm{L,M,T}} < m^* \tag{18.17}$$

and generally, in mechanics:

$$m^*_{\mathrm{M}} < m^*_{\mathrm{T}} < m^*_{\mathrm{L}}. \tag{18.18}$$

The ratio of the number of quantities whose dimensional equations contain a certain fundamental quantity m_f^* to the sum of all these numbers corresponding to the fundamental quantities of the system of units adopted, Σm_f^*, defines the frequency ν_f^* of the fundamental quantity considered; this ratio is always smaller than unity, i.e.:

$$\frac{m_f^*}{\Sigma m_f^*} < 1. \tag{18.19}$$

It is evident that the free term of equation (9.2) expressing the dimensional homogeneity condition is always a number; it represents the dimensional exponent of the quantity made explicit. The sign, either positive or negative, of any term of equations (9.2) is given by the sign of the dimensional exponent.

From what has been said above, it follows that in equations (9.2), written to relate the dimensional to the quantity exponents, any zero dimensional exponent of a fundamental quantity will lead to the elimination of the corresponding term from the equation of that fundamental quantity. Owing to the fact that the fundamental quantities do not have the same bearing on the dimensional equations written for the physical quantities, the equations which express the dimensional homogeneity condition will have variable numbers of t_f^* terms. As a general rule and for the field of mechanics, equations (9.2) for the fundamental quantity L include more t_L^* terms than the same equations when written for the fundamental quantities M and T; the number of t_M^* terms from equations (9.2) written for the fundamental quantity M is smaller than the number of m_T^* terms from the same equations as written for the fundamental quantity T. Taking the case of the pressure drop through a pipe having a circular cross-section, for which the following relation can be written

$$\Delta p = k\gamma^a v^b l^c g^e d^h, \tag{18.7}$$

the dimensional equations for $d^* = 3$ will be derived from the following dimensional matrix:

	Δp	γ	v	l	g	d
L	−1	−2	1	1	1	1
M	1	1	0	0	0	0
T	−2	−2	−1	0	−2	0

(18.20)

We thus obtain :

$$m_L^* = 6,\ m_M^* = 2,\ m_T^* = 4. \tag{18.21}$$

The system of algebraic equations expressing the dimensional homogeneity condition is of the form :

$$\begin{aligned} -2a + b + c + e + d &= -1 \quad (L) \\ a &= 1 \quad (M) \\ -2a - b - 2e &= -2 \quad (T). \end{aligned} \tag{18.22}$$

It should be noted that numbers (18.21) represent at the same time the numbers of terms from equations (18.22).

We remark that for a given function of variables which describe a given physical phenomenon and a given system of units, the number of terms in the equations which express the dimensional homogeneity condition differ from one equation to another ; however, some exceptions are possible — **Remark 5**

As given, the simplest equation (9.2) must have at least two terms ; when one term refers to an explicit quantity, the equation considered is a determinate one and can be solved for the singular solution. Any equation with $n^* > 2$ is indeterminate ; such an equation might take for instance the form

$$ax + by = c. \tag{18.23}$$

In order to solve equation (18.23), an additional condition is required ; it follows therefore that the dimensional analysis has to handle equations with additional conditions, that is Diophantine equations.

18.5 Derivatives of a Non-Integer Order

As given in connection with the values of dimension exponents, in mathematics derivatives of a non-integer order were defined. Since in the current literature this problem is not dealt with, the author has deemed it necessary to give here a brief review of the definition of these derivatives.

According to J. Liouville [4.4], the first to venture any ideas on the derivatives of a non-integer order was Leibniz. Also concerned with this problem were Euler, Lacroix, Laplace and Fourier. Let y be a function of x, i.e.

$$y = f(x). \tag{18.24}$$

The series development of this function is

$$y = A_1 e^{m_1 x} + A_2 e^{m_2 x} + A_3 e^{m_3 x} + \cdots \tag{18.25}$$

Let us note

$$y = \Sigma A_m e^{mx}. \tag{18.26}$$

Liouville denotes *the μ-order derivative of function y* by $\frac{\mathrm{d}^\mu y}{\mathrm{d}x^\mu}$ and defines it in the form of the new function $\Sigma A_m e^{mx} m^\mu$ derived from y by multiplying each term of the development by the μ power of the corresponding exponent, m, i.e.

$$\frac{\mathrm{d}^\mu y}{\mathrm{d}x^\mu} = \Sigma A_m e^{mx} m^\mu. \tag{18.27}$$

This definition is valid for any μ, which can be – optionally – positive or negative, integral or rational, irrational or imaginary. A consequence of this definition is the fact that with derivatives of any order, the indices are summed up, that is

$$\frac{\mathrm{d}^\gamma \dfrac{\mathrm{d}^\mu F(x)}{\mathrm{d}x^\mu}}{\mathrm{d}x^\gamma} = \frac{\mathrm{d}^{\mu+\gamma} F(x)}{\mathrm{d}x^{\mu+\gamma}}. \tag{18.28}$$

Sometimes, Liouville calls the same new function $\Sigma A_m e^{mx} m^\mu$ as the integral of the y function of order $-\mu$ and writes

$$\int^{-\mu} y \mathrm{d}x^{-\mu} = \Sigma A_m e^{mx} m^{-\mu}. \tag{18.29}$$

As can be seen, the new notations are those suggested by Leibniz.

Liouville describes in his works the general theorems, which can be very easily demonstrated and are extremely useful. Furthermore, he shows how to calculate derivatives of a non-integer order for some simple functions.

To illustrate, let us consider the function

$$y = e^{3x} + e^{-x}. \tag{18.30}$$

The $\frac{1}{2}$ order derivative, or the $-\frac{1}{2}$ order integral is:

$$\frac{d^{\frac{1}{2}} y}{dx^{\frac{1}{2}}} = e^{3x}\sqrt{3} + e^{-x}\sqrt{-1}, \tag{18.31}$$

while the $-\frac{3}{2}$ order derivative or the $\frac{3}{2}$ order integral is

$$\frac{d^{-3/2} y}{dx^{-3/2}} = \frac{e^{3x}}{3\sqrt{3}} - \frac{e^{-x}}{\sqrt{1}}. \tag{18.32}$$

As can be seen, for $\mu = 1$ the derivative will be obtained according to the classic definition.

In the case of the function:

$$y = \frac{1}{1 - e^x} \tag{18.33}$$

the derivative of the non-integer order is obtained in two ways according to the sign of x. If $x < 0$, the convergent series may be written

$$\frac{1}{1 - e^x} = 1 + e^x + e^{2x} + e^{3x} + \ldots \tag{18.34}$$

and in such a case:

$$\frac{d^{\mu} \frac{1}{1 - e^x}}{dx^{\mu}} = e^x + e^{2x} \cdot 2^{\mu} + e^{3x} \cdot 3^{\mu} + \ldots \tag{18.35}$$

If $x > 0$, one can write

$$\frac{1}{1 - e^x} = -e^{-x} - e^{-2x} - e^{-3x} - \ldots \tag{18.36}$$

and in such a case:

$$\frac{\mathrm{d}^{\mu}\dfrac{1}{1-e^{x}}}{\mathrm{d}x^{\mu}} = -(-1)^{\mu}(e^{-x} + e^{-2x}\cdot 2^{\mu} + e^{-3x}\cdot 3^{\mu} + \ldots). \quad (18.37)$$

Obviously, these derivatives have no connexion with the definition relationships of the secondary quantities.

18.6 Systems of Triangular Equations

The systems of equations expressing the condition of dimensional homogeneity between the dimension and the quantity exponents all have an interesting characteristic: they have a variable number of terms and constitute a special type of system of equations, called triangular equations. Such a system of equations can be [4.5]:

— completely-triangular;

— truncated-triangular.

A system of linear equations is called completely-triangular if its equations can be written in such a sequence that:

— the first equation includes only one unknown;

— any other equation contains the unknowns (not necessarily all) from the preceding equation, plus a new one. The general form of a completely-triangular system of equations is:

$$\begin{aligned}
&a_{11}x_1 &&= d_1\\
&a_{21}x_1 + a_{22}x_2 &&= d_2\\
&a_{31}x_1 + a_{32}x_2 + a_{33}x_3 &&= d_3 \qquad (18.38)\\
&\cdots\cdots\cdots\cdots\cdots &&\\
&a_{n1}x_1 + a_{n2}x_2 + \ldots + a_{nn}x_n &&= d_n
\end{aligned}$$

where none of the diagonal coefficients, a_{11}, a_{22} ... a_{nn} is zero. In a completely-triangular system of equations, the number of equations is the same as that of the unknowns. Thus, for instance, the system of equations

$$\begin{aligned} x &= 1 \\ x - 2y &= 5 \\ x - 3y - z &= 9 \\ y + z - u &= 0 \end{aligned} \tag{18.39}$$

is completely-triangular.

A system of linear equations is called truncated-triangular if it is completely-triangular for any set of unknowns and includes in addition some unknowns which do not pertain to the set under consideration. The general form of a truncated-triangular system of equations is the following:

$$\begin{aligned} &c_{11}t_1 + c_{12}t_2 + \ldots + c_{1k}t_k + a_{11}x_1 &&= d_1 \\ &c_{21}t_1 + c_{22}t_2 + \ldots + c_{2k}t_k + a_{21}x_1 + a_{22}x_2 &&= d_2 \\ &\ldots\ldots\ldots\ldots\ldots\ldots\ldots\ldots \\ &c_{n1}t_1 + c_{n2}t_2 + \ldots + c_{nk}t_k + a_{n1}x_1 + \ldots + a_{nn}x_n &&= d_n \end{aligned} \tag{18.40}$$

where all diagonal coefficients $a_{11}, a_{22}, \ldots, a_{nn}$ differ from zero. In a truncated-triangular system of equations, the number of equations is lower than that of the unknowns. Thus, for instance, the system of equations

$$\begin{aligned} t + u + x &= 5 \\ 3t + u + x + 2y &= 9 \\ -u + z &= 0 \\ x - y + z + w &= 0 \end{aligned} \tag{18.41}$$

is a truncated-triangular one. This system of equations is completely-triangular with respect to unknowns x, y, z and w.

When solving a completely-triangular system of equations the first equation is considered; this equation admits the solution

$$x_1 = \frac{d_{11}}{a_{11}}. \tag{18.42}$$

Replacing solution (18.42) in the remaining of equations (18.38) results in :

$$
\begin{aligned}
a_{22}x_2 \qquad\qquad &= d_2 - a_{21}\frac{d_1}{a_{11}}. \\
a_{32}x_2 + a_{33}x_3 \qquad &= d_3 - a_{31}\frac{d_1}{a_{11}} \\
\ldots\ldots\ldots\ldots\ldots\ldots & \\
a_{n2}x_2 + \ldots + a_{nn}x_n &= d_n - a_{n1}\frac{d_1}{a_{11}}.
\end{aligned}
\tag{18.43}
$$

Now we solve the first equation of the above system of equations and introduce the solution in the remaining equations. If this operation is repeated n times, the singular solution of the system of equations will be found. This conclusion is the basis of the following theorem : *any completely-triangular system of equations admits a singular solution.*

For solving a truncated-triangular system of equations, (18.40), the system will be written in the form :

$$
\begin{aligned}
a_{11}x_1 \qquad\qquad &= d_1 - c_{11}t_1 - \ldots - c_{1k}t_k \\
a_{21}x_1 + a_{22}x_2 \qquad &= d_2 - c_{21}t_1 - \ldots - c_{2k}t_k \\
\ldots\ldots\ldots\ldots\ldots\ldots & \\
a_{n1}x_1 + a_{n2}x_2 + \ldots + a_{nn}x_n &= d_n - c_{n1}t_1 - \ldots - c_{nk}t_k.
\end{aligned}
\tag{18.44}
$$

Assuming the unknowns $t_1, t_2, \ldots, t_k$ take random values :

$$t_1 = t_1', \quad t_2 = t_2', \ldots, t_k = t_k' \tag{18.45}$$

a completely-triangular system of equations will be obtained admitting the singular solution :

$$x_1 = x_1', \quad x_2 = x_2', \ldots, x_n = x_n'. \tag{18.46}$$

It is thus obvious that a system of t_i values leads to a singular solution for the system of equations (18.40) and, consequently, this system of equations accepts an infinity of solutions. On the basis of this conclusion, the following theorem can be stated: *any truncated-triangular system of equations accepts an infinity of solutions.* To solve such a system, all that is needed is to solve the first equation with respect to x_1, that is

$$x_1 = \frac{d_1}{a_{11}} - \frac{c_{11}}{a_{11}} t_1 - \ldots - \frac{c_{1k}}{a_{11}} t_k, \tag{18.47}$$

and then to substitute this solution in all the other equations. Thus a new truncated triangular system of equations is obtained, having as unknowns $x_2, x_3, \ldots, x_n$ and $t_1, t_2, \ldots, t_k$. Then x_2 is determined from the new system of equations and another truncated-triangular system of equations is built up, having as unknowns $x_3, x_4, \ldots, x_n$ and $t_1, \ldots, t_k$. Eventually, the general solution is obtained where unknowns $x_1, x_2, \ldots, x_n$ are expressed through linear polynomials of unknowns $t_1, t_2, \ldots, t_k$; these unknowns can take values at random, out of a given system of numbers.

In the literature, other methods of resolution are also suggested for the triangular systems of equations, such as that suggested by Novoselov [4.5] [18.1]. It should be stressed that when solving a truncated-triangular system of equations, the equations should be solved separately, in the increasing order of the number of terms. The systems of equations between dimension and quantity exponents which are written in dimensional analysis are truncated-triangular systems of equations; however, exceptions can exist as well.

18.7 Range and Nature of Dimension and Quantity Exponents

At the present state of scientific knowledge, the values of the quantity exponents contained in the relations of definition of the physical quantities are small, being placed at the beginning of the natural sequence of numbers. The values of the dimension exponents in the dimensional equations of the physical quantities are derived from these values. In the International System of units (SI), for instance, the significant values of the dimension exponents are confined within a narrow range, whether integers or zero, positive or negative. This range is shown in Table 18.2 below.

It should be mentioned that there are also some exceptions; thus, the dimensional equation for the electric bi-refriaction —

[18.1] p. 262.

Kerr's constant – contains T^6, while the dimensional equation for the molecule polarizability includes T^4 [1.20][18.2].

In other systems of units, the dimensional equations have rational dimension exponents; their numerators and denominators have values comprised within the same limits which, from the dimensional analysis standpoint, do not result in any essential changes.

TABLE 18.2. **Range of the values of dimension exponents in the international system (SI)**

No. of order	Fundamental physical quantity	Symbol	Range of values
1	Length	L	−4 ... + 5
2	Mass	M	−1, + 1
3	Time	T	−3 ... + 1
4	Intensity of electrical current	I	−2 ... + 2
5	Thermodynamic temperature	Θ	−1, + 1
6	Intensitaty of light	J	+ 1

The condition of dimensional homogeneity of physical relations is expressed through the linear equations found between the dimension and the quantity exponents. The values of the quantity exponents are estimated from the values of the dimension exponents. Consequently: *the range and nature of values of the quantity exponents are, for all practical purposes, similar to those of the dimension exponents.*

19 General Dimensional Analysis (I)

19.1 Preliminary Distribution of Variables

Let us consider a dimensionally homogeneous function of m^* variables

$$f(A, B, C, \ldots, M, N, P, \ldots) = 0, \tag{3.2}$$

arising in any determinate physical phenomenon. Remark 2 (Section IV. 18.2), mentioned in connexion with the Π theorem and Rayleigh's method, underlined the possibility of writing non-

18.2 p. 116.

dimensional products of powers which could be erroneous as to the physical dependence between the variables. The direction of change of the dependent variables with respect to the independent ones is not taken into consideration when writing the non-dimensional products of powers. Nor is it taken into consideration when, in accordance with Rayleigh's method, one selects at random the unknowns for solving an indeterminate system of equations expressing the condition of dimensional homogeneity of a physical relationship. This sense of change is found by dimensional analysis.

In the author's opinion, the decision about the director reversed sense of change of the dependent variables as a function of each independent variable arising in a definite physical phenomenon is not a problem of dimensional analysis. This problem should be solved when the list of the variables is drawn up. Consequently, the relation used in dimensional analysis computations should be written, say, in the form

$$f_1(M, N, P, \ldots) = f_2(A, B, C, \ldots). \tag{19.1}$$

The sense of the variables dependence could be assessed either through the theory of the phenomenon, or experimentally.

Obviously, the preliminary distribution among the two members will have as a consequence the essential fact that the quantity exponents will take only positive values.

19.2 Unknown Exponents for all Variables

Let us consider the relation between those variables which describe a definite physical phenomenon, written in compliance with Rayleigh's method:

$$P = kA^a B^b C^c \ldots M^m N^n \tag{8.17}$$

Remark 4 (Section IV. 18.3) underlines the fact that by ascribing to P an explicit value – i.e. ascribing to the corresponding quantity exponent a unit value – a certain solution for the system of equations is individualized; this solution describes the condition of dimensional homogeneity of relation (8.17) for the remaining quantity exponents.

The author considers that the restriction currently imposed when writing the equations between the dimension and the quantity exponents, i.e. to explicitly define one quantity, should be eliminated.

This idea is based on the concept of implicated relations which could take, as given, a variety of forms; one of these implicated functions, different from the standard one, can be more easily determined. As a consequence, the dimensional analysis calculations should operate with unknown quantity exponents that will be ascribed to all variables of a given function; hence, the relation between variables will be written in the form

$$M^m N^n P^p = kA^a B^b C^c \ldots \tag{19.2}$$

Obviously, ascribing unknown exponents to all variables will have the important result that — with indeterminate systems of equations — only convenient values, say integers, can be ascribed to quantity exponents.

19.3 Theorem of General Dimensional Analysis

Relation (19.2) is the author's conclusion based on physical investigations and fundamental, original reasonings. In what follows the author will describe a mathematical procedure which helps to achieve the same result.

In compliance with the theorem of dimensional analysis (Section II.9.6), the relation between the variables that describe a given physical phenomenon is (8.17). The quantity exponents $a, b, c, \ldots$ can take only positive or negative, integral or rational (fractions of integers) values. As a consequence, the most general form of relation (8.17) is

$$P = k\,\frac{A^{\frac{a_0}{a_1}} B^{\frac{b_0}{b_1}} C^{\frac{c_0}{c_1}} \ldots}{M^{\frac{m_0}{m_1}} N^{\frac{n_0}{n_1}} \ldots}, \tag{19.3}$$

where quantity exponents are rational numbers; this hypothesis is the most general one because in some particular cases both rational and integral values, and sometimes even integral values only can be obtained.

As given, the multiplication of a rational number by a convenient integral number will yield an integral number. Multiplication can be operated by any multiple of the denominator of the rational number considered. Thus, for instance, multiplying the exponent of the quantity A by multiple a_1 results in

$$\frac{a_0}{a_1} \cdot \text{mult} \cdot a_1 = \text{mult} \cdot a_0 = a, \tag{19.4}$$

where a is a new symbol for the quantity exponent.

This particular characteristic can be extended to a sequence of rational numbers; in such a case, the convenient integer will be the common multiple of all denominators of the rational numbers considered. As a consequence, we can note

$$p = \text{mult} \cdot (a_1, b_1, \ldots, m_1, \ldots); \tag{19.5}$$

and the rational numbers from (19.3) become:

$$\begin{aligned} \frac{a_0}{a_1} \cdot \text{mult} \cdot (a_1, b_1, \ldots, m_1, \ldots) &= p\frac{a_0}{a_1} = a \\ \frac{b_0}{b_1} \cdot \text{mult} \cdot (a_1, b_1, \ldots, m_1, \ldots) &= p\frac{b_0}{b_1} = b \end{aligned} \tag{19.6}$$

where $a, b, c, \ldots$ are new symbols for the quantity exponents. From relations (19.6) we obtain:

$$\frac{a_0}{a_1} = \frac{a}{p}, \quad \frac{b_0}{b_1} = \frac{b}{p}, \ldots. \tag{19.7}$$

Substituting in (19.3) the relations (19.7) results in:

$$M^{\frac{m}{p}} N^{\frac{n}{p}} P \ldots = kA^{\frac{a}{p}} B^{\frac{b}{p}} C^{\frac{c}{p}} \ldots \tag{19.8}$$

or else the equivalent relation:

$$M^m N^n P^p \ldots = kA^a B^b C^c \ldots, \tag{19.2}$$

already known.

Based on the above, we may state that *if the function*

$$P = f(A, B, C, \ldots, M, N, \ldots) \tag{3.3}$$

is dimensionally homogeneous and if P has a dimensional equation, there will be a product of powers $A^a B^b C^c \ldots$ *whose dimensional equation is the same as that of the product of powers* $M^m N^n P^p \ldots$

19.4 Repeated Explicit Definition Using Various Variables

Let us consider a system of linear equations (9.2) expressing the condition of dimensional homogeneity of a physical relation. As already shown (Section IV.18.4), Remark 5 refers to the fact that the number of terms of equations (9.2) differ from one equation to another, which means that some coefficients will have a zero value. These systems of equations are of the truncated-triangular type, and consequently the equations will be separately solved in the increasing order of the number of terms [4.5][19.1]. The possible exceptions will be solved in the same way as in the general case.

In the author's opinion, one way of step-by-step solving a truncated-triangular system of equations in dimensional analysis is by explicitly defining various variables. The step-by-step resolution of the equations allows for the explicit definition of one quantity for each equation and the quantity so defined should differ from one equation to another.

Of course, making the quantity explicit means ascribing a unit value to a corresponding quantity exponent; this unit value will be the smallest from the system of positive, integral numbers, i.e. it will be minimum. The systems of equations which cannot be solved by only one explicit definition for each equation will be solved in a similar manner, namely minimum values will be adopted for several unknowns. But this problem will be considered elsewhere (Section IV.19.6).

19.5 Progressive Dimensional Homogenization of Physical Relationships

The step-by-step resolution of the systems of equations expressing the condition of dimensional homogeneity imposed on physical rela-

[19.1] p. 261.

tions has a special meaning in dimensional analysis; it shows how to build up a monomial relation on the basis of variables of a dimensionally-given homogeneous function.

Let us consider a dimensionally homogeneous function of m^* variables along with a system of units having $d^* = 3$ fundamental quantities. The number of quantities whose dimensional equations contain a certain fundamental quantity was noted by m_f^*. Let us assume that $m_L^* = 6$, $m_M^* = 2$ and $m_T^* = 4$, consequently, the frequency of the fundamental quantities will be $\nu_L^* = 0.50$, $\nu_M^* = 0.16$ and $\nu_T^* = 0.33$. Hence, $\nu_M^* < \nu_T^* < \nu_L^*$ and the order in which fundamental quantities will be homogenized is M, T, L.

This finding leads to the rule of the progressive dimensional homogenization of physical relationships which is stated thus: *any monomial relation between dimensional physical variables results from a progressive homogenization of fundamental quantities involved in the dimensional equations of variables and is operated in the increasing order of their frequency (recurrence).*

19.6 Positive, Integral, Minimum Quantity Exponents

Let us consider a dimensionally homogeneous function (3.2) describing any definite physical phenomenon. The dependent variable, say P, or any other relevant quantity, will be expressed, as a function of the other variables, through the relation

$$P = kA^aB^bC^c \dots M^mN^n \dots \tag{8.17}$$

The form of the above relation is imposed by physical considerations and all variables intervening in any physical phenomenon should be covered by the relation expressing the law of the phenomenon. The monomial form of the relation is required by the method of study which is based on the condition of dimensional homogeneity of the relation. This form of relation between variables may be established by restricted dimensional analysis, when $i^* = 1$, up to the value of the numerical factor. To this end, the dimensional equations of the variables will be written according to a system of units having d^* fundamental quantities; these relations result, for example, from the following dimensional matrix:

$$\begin{array}{c|cccc|} & A & B & C\ldots & P\ldots \\ \hline \mathrm{L} & l_a & l_b & l_c & l_p \\ \mathrm{M} & m_a & m_b & m_c & m_p \\ \mathrm{T} & t_a & t_b & t_c & t_p \end{array} \tag{9.1}$$

Dimensional analysis calculations are conducted according to the condition of dimensional homogeneity, which must be imposed on any physical relation. This condition is expressed by the linear equations between the dimension exponents $l_a, l_b, \ldots, t_p$ and the quantity exponents $a, b, c, \ldots p, \ldots$:

$$\begin{aligned} l_a \cdot a + l_b \cdot b + l_c \cdot c + \ldots + l_m \cdot m + l_n \cdot n + \ldots &= l_p \\ m_a \cdot a + m_b \cdot b + m_c \cdot c + \ldots + m_m \cdot m + m_n \cdot n + \ldots &= m_p \\ t_a \cdot a + t_b \cdot b + t_c \cdot c + \ldots + t_m \cdot m + t_n \cdot n + \ldots &= t_p. \end{aligned} \tag{9.2}$$

When searching for the convenient form of the relations, it should be pointed out that, through dimensional analysis, the functions of variables can be systematized in three distinct cases:

Case 1. This is the simplest. The procedure is the same as that currently used and the Π theorem is applied; thus, i^* non-dimensional products of powers are written, for example in the form:

$$\Pi_m = \frac{M}{A^{a_m} B^{b_m} C^{c_m}}, \quad \Pi_n = \frac{N}{A^{a_n} B^{b_n} C^{c_n}}, \ldots \tag{8.2}$$

In compliance with Rayleigh's method, relation (3.3) is written in the form

$$P = kA^a B^b C^c \ldots M^m N^n \ldots \tag{8.17}$$

to obtain the same number i^* of non-dimensional products of powers. The function of variables (3.2) is reduced to a relation of the form

$$\varphi(\Pi_m, \Pi_n, \ldots) = 0. \tag{8.1}$$

If $i^* = 1$, relation (10.1) becomes

$$\Pi = k \tag{11.1}$$

and its form will be determined up to the value of the numerical factor. If $i^* > 1$, the form of relation (8.1) cannot be determined by restricted dimensional analysis alone ; as known, in such situations experimental data or the theory of phenomena are used. The form of the relation between variables can therefore be determined by restricted dimensional analysis only in the case of homogeneous functions that obey the following condition : *the number of variables of the homogeneous function under consideration should be one unit higher than the number of distinct equations expressing the dimensional homogeneity condition of the relation.* As a consequence, dimensional analysis is operative in determining the relationships for physical phenomena described by four or five variables.

Case 2 is more complex : let us consider a physical phenomenon from mechanics, described by a function of $m^* = 6$ variables :

$$f(A, B, C, M, N, P) = 0. \tag{19.9}$$

By applying the already known procedures, this function will lead to a number of non-dimensional products of powers $i^* > 1$, which means that the relation between variables cannot be determined by dimensional analysis calculations alone. Indeed, written according to a system of units with $d^* = 3$ fundamental quantities, the dimensional equations of variables have, say, the form shown by the following dimensional matrix :

	A	B	C	M	N	P
L	l_a	l_b	l_c	l_m	l_n	l_p
M	0	m_b	0	0	0	m_p
T	t_a	t_b	0	t_m	0	t_p

(19.10)

from which it follows that some dimension exponents have a zero value. In the dimensional equations, different numbers of fundamental quantities are contained, i.e. $m^*_{\mathrm{L}} = 6$, $m^*_{\mathrm{M}} = 2$ and $m^*_{\mathrm{T}} = 4$. The fact that the sign of dimension exponents is overlooked simplifies the demonstration which follows without affecting the conclusions. Let us assume that between the dimension exponents, the following relations exist :

$$l_{m,n,p,a,b} = \text{mult} \cdot l_c \qquad (\text{L})$$

$$m_p = \text{mult} \cdot m_b \qquad (\text{M}) \tag{19.11}$$

$$t_{m,p,b} = \text{mult} \cdot t_a \qquad (\text{T}).$$

As already shown, in dimensional analysis, Remarks 1, 2, 3, 4, 5, as well as some algebraic considerations relating to the form of physical relations are worth keeping in mind.

It is necessary to determine, either on the basis of the theory of the physical phenomenon or the experimental data, the sense of variation of the dependent variable as a function of each of the independent variables ; relation (3.3) can be written, for instance, in the form

$$f_1(M, N, P) = f_2(A, B, C). \tag{19.12}$$

The aim is to determine, through dimensional analysis, the form of relation (19.12) ; the method followed is based on the condition of dimensional homogeneity, which should be adhered to by any physical relation and thus leads to

$$[f_1] = [f_2]. \tag{19.13}$$

In order to avoid the individualization of a certain solution by using a variable that has a unit quantity exponent in an explicit form, unknown quantity exponents should be ascribed to all variables.

The values of the quantity exponents result from the equations expressing the condition of dimensional homogeneity. Taking into account the fact that the system of equations between the dimension and the quantity exponents is a truncated-triangular one, the equations will be solved separately, in the increasing order of their number of terms. Likewise, we pay due consideration to the fact that these equations are indeterminate and, therefore, the solution of an indeterminate equation will be a solution in integral numbers [4.5][19.2]. Consequently, the equations can be separately solved

---

[19.2] p. 299.

by repeatedly putting various variables into an explicit form; this procedure is the standard one and yields one additional element, which helps to select the solution leading to the correct relation.

As seen from the dimensional matrix (19.10), the fundamental quantity M occurs the least number of times $m_M^* = 2$; relation (19.12) will be written as

$$P^p f_1(M, N) = kB^b f_2(A, C). \tag{19.14}$$

The dimensional equations of variables P and B are introduced in relation (19.14):

$$(\mathrm{L}^{l_p}\mathrm{M}^{m_p}\mathrm{T}^{t_p})^p f_1(M, N) = k(\mathrm{L}^{l_b}\mathrm{M}^{m_b}\mathrm{T}^{t_b})^b f_2(A, C). \tag{19.15}$$

The dimensional homogeneity condition for the fundamental quantity M is expressed by relation

$$m_p \cdot p = m_b \cdot b. \tag{19.16}$$

From relation (19.15), variable P is written in an explicit form, i.e.

$$p = p_1 = 1. \tag{19.17}$$

This value is positive — as a result of the preliminary distribution of variables, it is an integral — because the equation (19.16) is an indeterminate one, and is minimum since the quantity P was written in an explicit form. Equations (19.16) and (19.17) lead to the singular solution:

$$b = b_1 = \frac{m_p}{m_b}. \tag{19.18}$$

This solution is also positive, integral — on the basis of relation (19.11, M) — and minimum.

Let us now consider, in the same manner, the fundamental quantity T, for which $m_T^* = 4$; consequently, the relevant equation between the dimension and the quantity exponents will have a greater number of terms. Based on the result obtained above, the relation will be written in the form

$$M^m P f_1(N) = kA^a B^{b_1} f_2(C). \tag{19.19}$$

Introducing the dimensional equations of variables M, P, A and B results in :

$$(\mathrm{L}^{l_m}\mathrm{T}^{t_m})^m\,(\mathrm{L}^{l_p}\mathrm{M}^{m_p}\mathrm{T}^{t_p})f_1(N)=k(\mathrm{L}^{l_a}\mathrm{T}^{t_a})^a\,(\mathrm{L}^{l_b}\mathrm{M}^{m_b}\mathrm{T}^{t_b})^{b_1}f_2(C). \quad (19.20)$$

For the fundamental quantity T, the condition of dimensional homogeneity is expressed by equation :

$$t_m \cdot m + t_p = t_a \cdot a + t_b \cdot b_1. \quad (19.21)$$

From relation (19.20) the variable M is written in an explicit form :

$$m = m_1 = 1. \quad (19.22)$$

This value is again positive, integral and minimum. Equations (19.21) and (19.22) lead to the solution :

$$a = a_1 = \frac{t_m}{t_a} + \frac{t_p}{t_a} - b_1\frac{t_b}{t_a}. \quad (19.23)$$

Solution a_1 is positive, integral — on the basis of relation (19.11, T) — and minimum.

The fundamental quantity L is comprised in the dimensional equations of all variables, $m_L^* = 6$; it will therefore lead to an equation between dimension and quantity exponents that will have the greatest number of terms. As above, this relation will be written, say :

$$MN^nP = kA^{a_1}B^{b_1}C^c. \quad (19.24)$$

Introducing the dimensional equations of variables M, N, P, A, B and C results in :

$$(\mathrm{L}^{l_m}\mathrm{T}^{t_m})(\mathrm{L}^{l_n})^n\,(\mathrm{L}^{l_p}\mathrm{M}^{m_p}\mathrm{T}^{t_p}) = k(\mathrm{L}^{l_a}\mathrm{T}^{t_a})^{a_1}\,(\mathrm{L}^{l_b}\mathrm{M}^{m_b}\mathrm{T}^{t_b})^{b_l}\,(\mathrm{L}^{l_c})^c. \quad (19.25)$$

The condition of dimensional homogeneity for the fundamental quantity L is expressed by equation :

$$l_m + l_n \cdot n + l_p = l_a \cdot a_1 + l_b \cdot b_1 + l_c \cdot c \quad (19.26)$$

From relation (19.24), variable N is put in an explicit form, i.e.

$$n = n_1 = 1. \quad (19.27)$$

Obviously, in this case the value obtained is also positive, integral and minimum. Equations (19.26) and (19.27) lead to the unique solution :

$$c = c_1 = \frac{l_m}{l_c} + \frac{l_n}{l_c} + \frac{l_p}{l_c} - a_1 \frac{l_a}{l_c} - b_1 \frac{l_b}{l_c}; \tag{19.28}$$

the value of solution c_1 is positive, integral —on the basis of relation (19.11, L) — and minimum.

On the basis of values found for the quantity exponents, the relation between the variables takes the form

$$P = k \frac{A^{a_1} B^{b_1} C^{c_1}}{M^{m_1} N^{n_1}} \tag{19.29}$$

which means that we obtained $i^* = 1$.

Case 3 is the general case, the function (3.2) containing any number of variables; furthermore, no conditions/or restrictions are imposed on the number of fundamental quantities and/or the dimensional equations of variables. In such a case, the number of non-dimensional products of powers will be $i^* > 1$. Obviously, the equations expressing the condition of dimensional homogeneity will differ from those used in Case 2, when the equations between the dimension and the quantity exponents permitted a singular solution to be reached by putting them into an explicit form. One should also bear in mind the fact that the number of unknowns could rise while that of the possible definitions could either remain the same or rise at a slower rate; in such a situation, the values of all quantity exponents could no longer be determined. The explanation can be found in the correlation

$$f(m^*, d^*) = 0. \tag{19.30}$$

However, it is obvious that the connexion between m^* and d^* can in no way affect the condition according to which the quantity exponents should be chosen out of the system of positive, integral, minimum numbers. It was also shown in this work that implicated relations can be written, which differ one from another only in form; furthermore, in a set of implicated relations, any relation can be the correct one for the given physical phenomenon described by the said relation. In such a situation one can always impose the condition that the quantity exponents should be positive, integral and minimum in value.

Should we have had available a convenient system of units, we could also have — by repeatedly using explicit forms with various variables — established the singular values of the quantity exponents in the general case.

The procedure of putting the quantities into an explicit form by using various variables, as applied before, led to correct results since in the first equation only two unknowns were involved and, on the basis of the solution derived, the second equation was transformed into another one which still had only two unknowns; finally, the same number of unknowns were also included in the last equation. In general, however, putting a quantity into an explicit form cannot lead to singular solutions for the equations, since relations (19.11) are not always valid and the equations can have — originally or subsequent to the introduction of the previous solutions — more than two unknowns.

It was shown before (Section I.2.4) that the systems of units used in science contain various numbers of fundamental quantities. These systems are not always convenient for the determination of the values of quantity exponents by dimensional analysis. One should therefore have a particular method for solving the systems of equations expressing the condition of dimensional homogeneity.

Let us consider a system of equations (9.2) expressing the condition of dimensional homogeneity of a monomial relation. As given, with truncated-triangular systems (9.2) the equations are solved separately in the increasing order of the number of terms. The unknowns take values pertaining to the system of positive, integral numbers and are minimal.

The system of equations has, for example, the following form:

$$l_a \cdot a - l_a \cdot b + l_c \cdot c + l_m \cdot m - l_n \cdot n + l_p \cdot p = 0 \quad \text{(L)}$$

$$m_a \cdot a - m_p \cdot p = 0 \quad \text{(M)} \quad (19.31)$$

$$t_a \cdot a + t_c \cdot c - t_p \cdot p = 0 \quad \text{(T).}$$

It is a system of equations with additional conditions, that is a system of Diophantine equations; the unknowns take only values which are positive — as a result of the preliminary distribution of the variables between the two terms of the monomial relation: integral — due to the fact that by ascribing unknown exponents to all quantities the resulting equations are indeterminate; and minimum — according to the reasoning concerning both the repeated explicit definition with various variables and the possibility of writing implicated relations.

To solve the system, let us first consider the equation (19.31, M) that has the smallest number of terms. This homogeneous equation is of the form:

$$ax + by = c, \tag{19.32}$$

for $c = 0$. To solve an indeterminate linear equation of the form (19.32) in integral numbers, we apply the algebraic theorem: *let a and b be two primes to each other, while x_0 and y_0, are a solution in integers for equation* (19—32); *in this case the relations* are:

$$x = x_0 - bi, \quad y = y_0 + ai \tag{19.33}$$

where $i = 0, 1, 2, 3, 4, \ldots$, yield all the solutions of equation(19.32). The theorem applies also in the case of a homogeneous equation for which $c = 0$ [4.2][19.3]; in such a case, relations (19.33) become:

$$x = -bi, \quad y = ai. \tag{19.34}$$

Equation (19.31, M) is solved for, say, a:

$$a = \frac{m_p}{m_a} p \cdot \tag{19.35}$$

Obviously, a can have an integral value only if

$$p = \text{mult. } m_a. \tag{19.36}$$

Any integer p, which is a multiple of m_a can be put in the form

$$p = m_a \cdot i, \tag{19.37}$$

where i can take the values of the natural sequence of numbers. The condition requiring that p be minimal leads to:

$$i = 1 \tag{19.38}$$

[19.3] p. 8 through 11.

that is the value taken into consideration when putting a variable into an explicit form. It follows from relations (19.37) and (19.38) that

$$p = p_1 = m_a, \tag{19.39}$$

while from relations (19.35) and (19.39) it follows that

$$a = a_1 = m_p. \tag{19.40}$$

So, the minimum values of quantity exponents correspond to the dimension exponents.

Introducing in equation (19.31, T), which has one more term, values a_1 and p_1 will lead to

$$t_a \cdot a_1 + t_c \cdot c - t_p \cdot p_1 = 0 \tag{19.41}$$

and hence to the singular solution:

$$c = c_1 = - a_1 \frac{t_a}{t_c} + p_1 \frac{t_p}{t_c} . \tag{19.42}$$

The condition for c to take an integral value is

$$- t_a \cdot a_1 + t_p \cdot p_1 = \text{mult} \cdot t_c. \tag{19.43}$$

If values a_1 and p_1 do not satisfy relation (19.43), we shall take $i = 2, 3, \ldots$ and resume the calculations. However, it is generally found that in practice such trials are not necessary. We shall assume that the c_1 value obtained is an integer.

Equation (19.31, L), with the greatest number of terms, takes the form:

$$l_a \cdot a_1 - l_b \cdot b + l_c \cdot c + l_m \cdot m - l_n \cdot n + l_p \cdot p_1 = 0 \tag{19.44}$$

where b, m and n are unknowns, which are separated thus:

$$l_b \cdot b - l_m \cdot m + l_n \cdot n = l_a \cdot a_1 + l_c \cdot c_1 + l_p \cdot p_1. \tag{19.45}$$

The right-hand member of equation (19.45) has a known value v. Let us proceed with the following substitution:

$$l_b \cdot b - l_m \cdot m = x. \tag{19.46}$$

Equation (19.38) will take the form

$$x + l_n \cdot n = v. \tag{19.47}$$

We choose two positive, minimum, integral values:

$$x = x_1, \quad n = n_1 \tag{19.48}$$

which satisfy the relation (19.48); this choice does not raise any difficulties. It follows immediately from equations (19.46) and (19.48) that

$$l_b \cdot b - l_m \cdot m = x_1. \tag{19.49}$$

For solving the above equation in integral numbers, the same theorem as before is used; in such a case:

$$b = b_1 - l_m \cdot i, \qquad m = m_1 + l_b \cdot i \tag{19.50}$$

where $i = 0, 1, 2, 3, 4, \ldots$, yield all the solutions of equation (19.49). Of course, the minimum solutions will result for

$$i = 1. \tag{19.51}$$

When rational values are obtained, higher values will be chosen for x_1 and n_1, corresponding to $i = 2, 3, \ldots$ and the calculations will be resumed. In practice, usually at most two or three trials are needed.

Using this method, the singular values of all quantity exponents can be determined for the general case as well, which means that $i^* = 1$. In practice however, sometimes, for exceptional cases, two or even three solutions — ambiguity — having positive, integral and minimum values could be obtained.

By changing the system of units, one can demonstrate that the unit values adopted when using a certain d^* correspond to the values of exponents of those quantities which were put into an explicit form when a higher d^* is used.

To prevent any doubt about the validity of the integral quantity exponents of the relations describing the physical phenomena, it should be emphasized that this method is a dimensional one and the relations obtained are implicated forms of relations known usually in a different form.

The demonstration above leads to the following statement: *any accurate, dimensionally homogeneous physical relation the form of which is a product of powers accepts positive, integral, minimum quantity exponents.*

This will be dealt with elsewhere in more detail (Section IV. 20.2).

19.7 Solving Diophantine Systems of Equations (I)

In general dimensional analysis, the relation between variables describing a definite physical phenomenon is written in the form:

$$M^m N^n P^p \ldots = k\, A^a B^b C^c \ldots . \tag{19.2}$$

In those systems of homogeneous equations that express the condition of dimensional homogeneity of a monomial physical relation, the unknowns represent the quantity exponents of variables concerned with a definite physical phenomenon. These are truncated-triangular systems of equations and if $i^* > 1$, which is precisely the object of the general dimensional analysis, they are always indeterminate. The rule (I) to follow in solving Diophantine systems of equations can be stated thus: *any homogeneous system of linear equations expressing the condition of dimensional homogeneity of an accurate physical relation is solved by steps: the equations are taken in the increasing order of their number of terms and positive, integral, minimum values are ascribed to the independent unknowns so as to obtain similar values for the dependent ones.*

In this book, whenever deemed adequate, applications of this rule are presented.

19.8 Comparison of Systems of Equations

The systems of equations concerning both the dimension and the quantity exponents written for the restricted dimensional analysis differ from those written for general dimensional analysis.

In compliance with Rayleigh's method, the relation between variables is written

$$P = kA^a B^b C^c \ldots M^m N^n \ldots . \tag{8.17}$$

The condition of dimensional homogeneity for the above relation is expressed through a non-homogeneous system of equations

$$\sum_{i=a}^{m} x_i \cdot i = x_p, \tag{19.52}$$

including $e^* \leqslant d^*$ linear equations with $n^* = m^* - 1$ unknowns; symbol x stands for the dimension exponent while equations (19.52) are written for all fundamental quantities. Hence, equations (19.52) will be of the form:

$$\Sigma v_1 v_2 = k \tag{19.53}$$

where v_1 and k stand for the values of the dimension exponents, and v_2, for those of the quantity exponents. The condition of dimensional homogeneity is always expressed by systems of linear equations.

According to what was said above, the relation between the variables is written:

$$M^m N^n P^p \ldots = k A^a B^b C^c \ldots . \tag{19.2}$$

The condition of dimensional homogeneity of relation (19.2) will be expressed by a homogeneous system of equations:

$$\sum_{i=m}^{z} x \cdot i = \sum_{i=a}^{l} x_i \cdot i \tag{19.54}$$

built up with $e^* \leqslant d^*$ linear equations with $n^* = m^*$ unknowns. The number of unknowns is therefore one unit higher than in the restricted dimensional analysis. Hence, equations (19.54) will have the form

$$\Sigma v_1 v_2 = 0, \tag{19.55}$$

different therefore, from the equations currently used in dimensional analysis. The condition of dimensional homogeneity is always expressed by Diophantine systems of equations.

19.9 The Number of Non-dimensional Products of Powers

As known, the statement of the Π theorem also concerns the number of non-dimensional products of powers which may result from a dimensionally homogeneous function of m^* variables; this number, i^*, is given by the relation:

$$i^* = m^* - r^* \tag{8.8}$$

where r^* is the rank of the dimensional matrix of variables (Section II.8.1). The dimensional analysis calculations lead to the relation connecting the variables which describe a physical phenomenon – up to the value of the numerical factor – only when $i^* = 1$. For those situations when $i^* > 1$, the dimensional analysis alone cannot determine the form of the relation and use should be made of data supplied by experiment or, sometimes, by the theory of the phenomenon, to obtain the values of the exponents of non-dimensional products of powers.

As shown before, in some cases of indetermination, i.e. $i^* > 1$, it is still possible to obtain relations between the variables describing physical phenomena; this can be achieved by a preliminary distribution of variables, by ascribing unknown quantity exponents to all variables and by repeatedly putting into explicit form different variables. Moreover, through the last process one can find the values of a number of at most $d^* - 1$ quantity exponents. As a consequence, the number of non-dimensional products of powers will be given by the relation:

$$i^* \geqslant m^* - r^* - d^* + 1; \tag{19.56}$$

obviously, the above relation could lead to $i^* = 1$. Consequently, in some cases, by using either the Π theorem or Rayleigh's method, one can obtain relations between non-dimensional products of powers, while by using the method described by the author, relations between variables are obtained.

Likewise, it was already shown that for all cases of indetermination, $i^* > 1$, by using the preliminary distribution of variables, by ascribing unknown quantity exponents to all variables, and by applying the condition of the minimum values for quantity exponents, one can obtain relations between variables which describe physical phenomena. Moreover, by applying the rule about the solution of

Diophantine systems of equations one can determine the values of a number of $j^* - 1$ quantity exponents; this number depends on d^*, i.e.

$$j^* = f(d^*) \tag{19.57}$$

and always leads to:

$$i^* = m^* - r^* - j^* + 1 = 1, \tag{19.58}$$

that is : *in general dimensional analysis, the number of non-dimensional products of powers is always equal to unity.*

In compliance with the method of general dimensional analysis suggested in this book, namely to ascribe unknown exponents to all variables of a given function, let us write the relation in the form (19.2). In order to write the dimensional equations of the variables, we shall use the dimensional matrix (7.12). The dimensional homogeneity condition of relation (19.2) is expressed by the system of linear equations :

$$\begin{aligned} l_a \cdot a + l_b \cdot b + l_c \cdot c + \ldots &= l_m \cdot m + l_n \cdot n + l_p \cdot p + \ldots \\ m_a \cdot a + m_b \cdot b + m_c \cdot c + \ldots &= m_m \cdot m + m_n \cdot n + m_p \cdot p + \ldots \\ t_a \cdot a + t_b \cdot b + t_c \cdot c + \ldots &= t_m \cdot m + t_n \cdot n + t_p \cdot p + \ldots \end{aligned} \tag{19.59}$$

which is a homogeneous one. The number e^* of distinct (19.59) equations that can be written for a given, well-determined physical phenomenon may be at most equal to the number d^*, of fundamental quantities that build up the system of units chosen at random, that is

$$e^* \leqslant d^*; \tag{10.4}$$

which is a situation similar to that encountered in the restricted dimensional analysis. The number n^* of unknowns in equations (19.59) is always equal to the number of quantities used to describe the phenomenon under study. It follows therefore that

$$n^* = m^* \tag{19.60}$$

which differs from the situation existing in the restricted dimensional analysis, where $n^* = m^* - 1$.

For solving the systems of equations (19.59) use is made of the rule applied for the solution of Diophantine systems of equations. The result of calculations conducted according to dimensional analysis rules will be a monomial relation of the form:

$$P = k \frac{A^{a_1} B^{b_1} C^{c_1} \dots}{M^{m_1} N^{n_1} \dots} \tag{19.61}$$

which means that $i^* = 1$.

Some examples of general dimensional analysis applications are presented later in this book.

20 General Dimensional Analysis (II)

20.1 Function of Physical Variables

The subsequent stages of the study of a physical phenomenon by applying the general dimensional analysis are as follows:

— the determination of variables, preliminary distribution of the variables between the two members of the monomial relation and the assignment of unknown exponents to all variables;

— the general dimensional analysis calculations;

— the determination of the numerical factor from experimental data.

These stages differ from those known from restricted dimensional analysis in that:

— the preliminary distribution of variables and the assignment of unknown exponents to all variables are not performed in the case of restricted dimensional analysis;

— in the restricted dimensional analysis, the values of the exponents of non-dimensional products of powers are determined from experimental data or from the theory of the phenomenon, while in the general dimensional analysis these values are determined through dimensional analysis calculations for all quantity exponents.

Obviously, the result of dimensional analysis calculations depends on the variables function. Here, the following question might be asked: can any function of different variables accurately describe a certain physical phenomenon? The author has already shown that the listing of variables raises some difficulties and only a few general rules are available, as yet (Section III.12.2); this situation made Bridgman say that dimensional analysis should be applied only by expert physicists. This statement refers, of course, only to the first stage of the study and only when the dimensional methods are applied in order to derive new relations; dimensional analysis calculations are, in fact, easy. But general dimensional analysis also requires a preliminary distribution of variables; this *effort* has an important *effect* since it allows the values of all quantity exponents to be determined through dimensional analysis calculations alone.

A survey of the functions of variables used in dimensional analysis reveals the fact that in some lists foreign variables are included or else some of the required variables are missing; this is due to a poor knowledge of either the phenomenon under consideration or the other relations existing between some of the listed variables. It may happen that some variables are deliberately eliminated; this is the case when a characteristic variable is retained instead of several variables of like nature. It may also happen that some variables are ignored on account of their insignificant effects and of their constant values. Sometimes, the list includes variables that are covered by the common form of the relation or the variables of an implicated relation. From the dimensional analysis standpoint, any function of variables may be *accurate, equivalent accurate, approximate, or erroneous.* Any accurate function can be converted into an equivalent accurate function through substitution of one or more variables, on the basis of either the definition relations of physical quantities, or other known physical relations.

Any function of physical quantities describes a certain physical phenomenon. Sometimes, a certain function might describe a class of physical phenomena; take for example the physical phenomenon of a beam which is bent under the action of a force: the beam can be either supported or built in at both ends; or else it can be built in at one end and either supported or free at the other (four alternatives). Hence, the representability of the function of variables can be either simple or multiple; in the latter case, the differentiation is made on the basis of peculiarities of physical systems which, in the analytical method of study, are represented by the singularity — initial and limit — conditions.

Likewise, the preliminary distribution of variables between the two members of a monomial physical relation can be either correct or erroneous.

At the same time, the results of calculations conducted as to the general dimensional analysis are conditioned by the number of fundamental quantities of the system of units used ; a function of physical variables may or may not be dimensionally homogeneous for the fundamental quantities under consideration.

The nature of the function of variables is at the same time correlated to both the quality of the preliminary distribution of variables between the two members of the monomial relation and to the option regarding the number of fundamental quantities on which the dimensional relations of variables rely. Considering any function of variables, it is doubtful that one can tell from the very beginning, whether the function is correct or erroneous ; moreover, there is no way of being absolutely sure about either the quality of the preliminary distribution of variables or the correct choice of the set of fundamental quantities.

As a consequence, any dimensional analysis calculation must be subsequently confirmed, and this is a *sine qua non* condition. The basic criterion in establishing the validity of the results of the dimensional analysis calculations is only the practice, namely : experimental laboratory data or pilot plant data. A comparison criterion between the data computed by means of a relation derived through dimensional analysis and those supplied by measurements can be the numerical factor of proportionality ; this factor has a constant value.

To obtain complete results subsequent to the dimensional study of a physical phenomenon, experimental data should be available for the whole range of definition of the phenomenon under consideration. Data computed by means of relations derived through dimensional analysis should agree with those supplied by measurements.

It is the opinion of the author that any physical phenomenon or process can be described as an accurate function of variables.

20.2 Positive, Integral, Small Quantity Exponents

Inasmuch as the function of variables describing a certain determinate physical phenomenon might be accurate, equivalent-accurate, approximate or erroneous, the following question arises :

do the quantity exponents accept, for any function of variables, positive, integral, minimum values? The answer to this question will be given below.

Let us consider the dimensionally homogeneous function of m^* variables:

$$f(A, B, C, \ldots, M, N, P, \ldots) = 0. \tag{3.2}$$

It is assumed that one knows the sense of change of the dependent variable as a function of each of the independent variables. Let us further write a monomial relation having the variables distributed between the two members while the quantity exponents are unknown for all variables. The condition of dimensional homogeneity is expressed through a truncated-triangular system of equations, the solution of which relies on the rule applied for solving the system of Diophantine equations. The quantity exponents take positive, integral, minimum values; a monomial relation is obtained which is determined up to the numerical factor. It is further assumed that from the comparison with data supplied by experimental measurements, the relation leads to k = constant. For the sake of simplification, we shall consider a function of only $m^* = 4$ variables:

$$f(A, B, M, P) = 0 \tag{20.1}$$

the relation between variables being of the form

$$P = k \frac{A^{a_1} B^{b_1}}{M^{m_1}} \tag{20.2}$$

where $m^* = 4$ variables arise; this particular assumption does not affect the results.

It is assumed that out of the m^* variables a number of $m_1^* < m^*$ are connected through a relation of definition or by any other known physical relation. Thus it is admitted that

$$B = k \frac{C^\gamma}{N^\eta} \tag{20.3}$$

where the new variables C and N are included. By substituting relation (20.3) in (20.1), we obtain:

$$P = k \frac{A^{a_1} C^{b_1 \gamma}}{M^{m_1} N^{b_1 \eta}}, \tag{20.4}$$

which is equivalent to relation (20.2).

The new function of $m^* = 5$ variables

$$f(A,\ C,\ M,\ N,\ P) = 0 \tag{20.5}$$

is also dimensionally homogeneous. General analysis calculations lead to the relation between the variables of the function (20.5); the result may be of the form

$$P = k\,\frac{A^{a_1'}C^{c_1'}}{M^{m_1'}N^{n_1'}}\,. \tag{20.6}$$

The following question arises: is relation (20.4) identical to relation (20.6)? Of course, "identity" refers to the values of the quantity exponents. The answer will be determined by the value of the quantity exponent b_1 and by the form of relation (20.3). There can exist four alternatives.

Case 1: the quantity exponent has a unit value,

$$b_1 = 1 \tag{20.7}$$

while the quantity exponents γ and η have integral values. Hence, products $b_1\eta$ and $b_1\gamma$ will take integral values too:

$$b_1\gamma = \gamma, \qquad b_1\eta = \eta. \tag{20.8}$$

Consequently, relation (20.4) has the form

$$P = k\,\frac{A^{a_1}C^{\gamma}}{M^{m_1}N^{\eta}}\,. \tag{20.9}$$

This relation can be established through the general dimensional analysis and is identical with relation (20.6).

Case 2: the quantity exponent is

$$b_1 \neq 1, \tag{20.10}$$

while the quantity exponents γ and η have integral values; products $b_1\gamma$ and $b_1\eta$ will therefore take integral values too, and consequently the situation will be similar to that of case 1.

Case 3: the quantity exponent is

$$b_1 = 1, \tag{20.7}$$

while the quantity exponents γ and η have rational values (irreducible fractions); consequently, products $b_1\gamma$ and $b_1\eta$ will have rational values:

$$b_1\gamma = \frac{\gamma_0}{\gamma_1}, \qquad b_1\eta = \frac{\eta_0}{\eta_1} \tag{20.11}$$

and relation (20.4) will take the form

$$P = k\,\frac{A^{a_1}C^{\gamma_0/\gamma_1}}{M^{m_1}N^{\eta_0/\eta_1}}. \tag{20.12}$$

Obviously, this relation cannot be determined by means of the general dimensional analysis and of the theorem of positive, integral, minimum quantity exponents; it implies, however, another relation, whose form could be, say:

$$P^u = k\,\frac{A^{a_1 u}C^{\gamma_0/\gamma_1 \cdot u}}{M^{m_1}N^{\eta_0/\eta_1 \cdot u}} \tag{20.13}$$

where

$$u = \text{l.c.m.}\ (\gamma_1,\ \eta_1) \tag{20.14}$$

or

$$u = \text{mult}\cdot(\gamma_1,\ \eta_1). \tag{20.15}$$

The quantity exponents from relation (20.13) take integral values. For a u value, as obtained from relation (20.14), the quantity exponents from relation (20.13) have the least integral values, hence the relation is the simplest. On the other hand, relation (20.13) can be determined by general dimensional analysis only if the equations expressing the condition of dimensional homogeneity for values different from the minimum one can be solved. In solving these equations one shall take $i = 2, 3, \ldots$ Practice has shown that at most two or three trials suffice.

Case 4: the quantity exponent is

$$b_1 \neq 1, \tag{20.10}$$

while the quantity exponents γ and η take rational values. Products $b_1\gamma$ and $b_1\eta$ can take integral, integral and rational, or rational values. Consequently we are not confronted with a new situation.

From above, it can be seen that the quantity exponents take positive, integral, minimum values only for the accurate function. When the function is equivalent-accurate, the quantity exponent takes positive, integral, small values. This conclusion leads to the following statement: *any dimensionally homogeneous, equivalent-accurate physical relation under the form of a product of powers accepts positive, integral, small values for quantity exponents.*

The approximate function is not an accurate one. From the dimensional analysis standpoint, the approximate function is similar to the equivalent accurate function, since it includes variables that differ from those of the accurate function. As a consequence, for the approximate function, quantity exponents having positive, integral, small values shall be considered.

20.3 Solving Diophantine Systems of Equations (II)

From what was said above, it follows that the unknowns included in those systems of equations which express the condition of dimensional homogeneity of a physical relation written under the form

$$M^m N^n P^p \ldots = k A^a B^b C^c \ldots, \qquad (19.45)$$

take positive, integral, minimum values only when certain conditions — as described above — were adhered to. In all other cases, the unknowns take positive, integral, small values. Speaking about the small values of the quantity exponents, it was shown before (Section IV.18.8) that they are similar to the values of dimensional exponents; these values were listed in Table 18.2. This conclusion leads to the rule (II) for solving Diophantine systems of equations, namely: *any system of homogeneous linear equations expressing the condition of dimensional homogeneity of a physical relation should be solved in two stages: first the equations are considered in the increasing order of terms, then positive, integral, small values are ascribed to the independent unknowns, thus obtaining similar values for the dependent unknowns.* This rule is valid for those cases when the positive, integral, minimum values do not lead to correct physical relations. With systems of equations having several unknowns, it is sometimes better to give a systemized solution, i.e. to draw up tables of solutions so as to avoid any omission.

Some applications of this rule are presented in this book.

20.4 Theorem of Quantity Exponent Values

The demonstrations above have led to conclusions which can be summed up in one statement, namely *the dimensionally homogeneous, monomial physical relations can be determined with positive, integral, minimum or small quantity exponents*. This can be called the theorem of the quantity exponent values or the theorem of the positive, integral, minimum or small quantity exponents.

The study of physical phenomena through the dimensional method shall comply with the general dimensional analysis as described in this work. In solving the systems of equations including both the dimension and the quantity exponents, expressing the condition of dimensional homogeneity of a physical relation, the rule for solving the Diophantine systems of equations should be applied. This rule can be stated thus: *the systems of homogeneous linear equations expressing the condition of dimensional homogeneity of monomial physical relations shall be solved by stages: first the equations shall be considered in the increasing order of their number of terms, then positive, integral, minimum or small values shall be ascribed to the independent unknowns, thus obtaining for the dependent unknowns values of like nature.*

This rule shall be also applied to those exceptional cases when the systems of equations are not truncated-triangular ones.

The solution of systems of indeterminate equations expressing the condition of dimensional homogeneity should include only positive, integral values for all unknowns; in some cases, for some of the unknowns, the values adopted shall be immediately higher than unity.

The final result described above constitutes the heuristic procedure which is useful in practice.

In some cases, the quantity functions originally adopted can lead to correct relations having values different from the minimum ones. It was found however that by changing the quantity functions, minimum values lead – even in such cases – to correct relations.

Theoretically, in the case of an indeterminate system of equations expressing the dimensional homogeneity of a physical relation, the unknowns can take any integer of rational value, provided however that the relation degree is kept the same, and this is a *sine qua non* condition. This condition should be, of course, applied to all equations since they are resolved separately and in a progressive manner. The monomial physical relation which will be obtained shall have an implicated form; by extracting the root of the same order as that adopted when solving the equations, a relation identical with that obtained by observing the condition of minimum values for the quantity exponents will result. The method is correct but it involves a great many calculations because the solutions have to be checked.

In physics, the minimum values are ascribed to the quantity exponents of dimensionally homogeneous monomial relations. They are used in other fields too: in chemical equations, for the substance coefficients; and in nuclear reactions, for the coefficients of the nucleus and of the elemental particle.

A more comprehensive list of elements of physics, chemistry and nuclear physics which are compared in the light of the present book is given in Appendix 3, to stress the consistency of the three fields of science.

21 The Numerical Factor in the General Dimensional Analysis

21.1 The New Situation

The dimensional analysis calculations are always carried out for dimensionally homogeneous functions of m^* measurable variables. They rely on the condition of dimensional homogeneity, a condition to be observed by any physical relation, for in this way the relation validity is not lost with the change of the units of measurement. The dimensional analysis calculations lead to a relation whose form is that of a product of powers of variables. Any relation that describes a physical phenomenon includes physical quantities, universal constants and characteristic constants.

Relations derived through general dimensional analysis cannot differ from those derived through the analytic method. However, in most cases the latter relations also include a numerical factor the value of which is significant, namely

$$k = \text{const.} \neq 0. \tag{21.1}$$

As a consequence, whenever a relation is determined through general dimensional analysis, one should make sure that a numerical factor such as the one shown in relation (21.1) is not needed; the value of this factor would be

$$k \neq 1. \tag{21.2}$$

The numerical factor of proportionality is essential in determining physical relations through general dimensional analysis, because it is the link between the relation derived and the data found for the relevant variables by measurements carried out in laboratory conditions or in actual practice. Of course, the measured value of a dependent variable should be identical — within the accuracy boundaries permitted by the measuring and mathematical means available — with the value estimated with the relation derived through dimensional analysis. In fact, this factor correlates the value of the dependent variable with the value of the monomial built up with the various powers of the independent variables. Thus the relation

$$P = f(A, B, C, \ldots, M, N, \ldots) \tag{3.3}$$

is written in the form

$$P = kA^a B^b C^c \ldots M^m N^n \ldots, \tag{8.17}$$

where k is the numerical factor of proportionality.

From what was said above, one can state that *any physical monomial relation derived on the basis of dimensional homogeneity is correct if, and only if, the numerical factor of proportionality has a constant value.*

We can further state that:

— any physical relation derived through dimensional analysis calculations should contain a numerical factor which sometimes takes the unit value;

— the value of this numerical factor will be found by using experimental data, which means that the method applied to study the phenomenon includes practical application;

— the value of the numerical factor should be constant, which will confirm the validity of the elements used and of the deductions performed.

However, the general dimensional analysis introduces new considerations regarding the numerical factor; these considerations may refer to such aspects as:

— the correctness and representability of the function of variables established for the physical phenomenon under study;

— the correctness with which the variables were distributed between the members of the monomial relation, the dimensional homogeneity of the relation concerning the fundamental quantities selected, the validity of the small values ascribed to the quantity exponents.

Whenever the numerical factor of proportionality is determined from a monomial relation derived through general dimensional analysis, theoretically one single set of values for the variables included in a relation is needed since one single unknown has to be determined. It is, however, strongly recommended that the value determined for a factor k be checked on several sets of experimental data. The value of the factor should be constant — or practically constant — for all sets of data used. Any relation derived through dimensional analysis that checks the data found by actual measurements is correct and can be used in practice.

21.2 Significance of the Numerical Factor

Whenever the general dimensional analysis is applied to the study of physical phenomena, it is desirable to use accurate functions of physical quantities. For the quantity exponents, positive, integral, minimum values are ascribed first. Calculations carried out on the basis of the relation thus established and on the measured data may lead to:

$$k = \text{const.}$$
$$k \neq \text{const.} \tag{21.3}$$

When $k = \text{const.}$ the relation is correct, while when $k \neq \text{const.}$ the investigation will be continued until a correct relation is obtained. The latter could be due to the following causes:

— the function of physical quantities is equivalent-accurate, or approximate;

— the distribution of physical quantities between the two members of the monomial relation is erroneous.

If, after eliminating the errors, we still find that $k \neq \text{const.}$ positive, integral, small values will be ascribed to the quantity exponents; 2 up to 4 such values are admitted. If, by comparing the values as computed on the basis of the new relations with the data reported by measurements, we still find that $k \neq \text{const.}$ it follows that either the physical phenomenon under study does not accept a monomial relation, or that the function of variables is erroneous.

If, by using positive, integral, small quantity exponents, two or several relations result leading to nearly the same values, the relation that best fits the measurement data will be derived by using the mean square error.

V. Applications of the General Dimensional Analysis

22 The New Method and Some Comparative Considerations

22.1 Information to be Obtained by General Dimensional Analysis

Can general dimensional analysis produce the values of quantity exponents when $i^* > 1$? The answer to this question is, in fact, the object of general dimensional analysis (Section IV.18.1).

The author has reached the conclusion that in order to extend the range of information that could be obtained through general dimensional analysis, it is necessary when conducting a study on a physical phenomenon through dimensional analysis to follow these steps :

— impart the variables between the two members of the monomial relation before writing the equations expressing the condition of dimensional homogeneity ;

— ascribe unknown quantity exponents to all variables ;

— put into an explicit form the various variables that describe a definite physical phenomenon, or else seek a relation implicated by the relation written in the standard form.

General dimensional analysis could then yield new information, namely :

— the relation between physical quantities when $i^* > 1$, up to the numerical factor.

— the rational relationships derived by other methods can be checked to discover the nature of the function of variables.

Therefore, general dimensional analysis can supply a useful heuristic procedure for practical purposes. A comparison with the information that can be derived through restricted dimensional analysis will show that in general dimensional analysis no indeterminate φ functions appear, as in the relationships between non-dimensional products of powers. As for the check that can be conducted on the physical relationships through general dimensional analysis, it is the author's opinion that it is important to know the nature of the function, since in the case of inaccurate functions the study of the physical phenomena involved is not complete.

22.2 Deriving the Physical Relations

Any dimensionally homogeneous function of m^* variables can be written, according to previous discussions in this work, in the form of a monomial relation between variables.

To illustrate how the calculus is developed in general dimensional analysis, let us consider the physical phenomenon of a sphere falling through a viscous medium, already discussed in Section III.13.1.

According to the analytical method, the equation describing the phenomenon is :

$$\frac{\pi}{6}\, d^3 g \rho - 3\pi d \eta v = m\, \frac{\mathrm{d}v}{\mathrm{d}t}, \tag{22.1}$$

where the following variables are involved : the sphere weight, the lifting force, the drag resistance and the inertial resistance. The symbols denote : sphere diameter d, the normal acceleration of free fall g, the apparent density ρ, the dynamic viscosity η, the velocity of fall v, the mass m and time t. As the body goes deeper into the viscous medium, the motion becomes uniform and the velocity approaches a constant value ; relation (22.1) takes the form

$$\frac{\pi}{6}\, d^3 g \rho - 3\pi d \eta v = 0. \tag{22.2}$$

It follows from relation (22.2) that the velocity of fall is

$$v = \frac{1}{18}\frac{d^2 g \rho}{\eta}. \tag{22.3}$$

The function admitted for the dimensional analysis calculus includes $m^* = 5$ variables :

$$f(d, g, \rho, \eta, v) = 0. \tag{13.3}$$

Let us consider the dimensional equations of variables written in the system of units with $d^* = 3$ fundamental quantities ; they result from the following dimensional matrix :

	d	g	ρ	η	v
L	1	1	−3	−1	1
M	0	0	1	1	0
T	0	−2	0	−1	−1

(13.4)

The relation between variables is written in the form

$$\eta^a v^b = k\, d^c g^e \rho^h \tag{22.4}$$

where unknown quantity exponents were ascribed to all variables.

The condition of dimensional homogeneity of relation (22.4) is expressed through the homogeneous system of linear equations

$$\begin{aligned} -a + b - c - e + 3h &= 0 \quad (\mathrm{L}) \\ a - h &= 0 \quad (\mathrm{M}) \\ -a - b + 2e &= 0 \quad (\mathrm{T}) \end{aligned} \tag{22.5}$$

with $e^* = 3$ equations and $n^* = 5$ unknowns ; this system of equations is consequently an indeterminate one.

The above system of equations is solved by applying the rule previously stated in Section IV.19.7. Considering the equation (22.5,M), which has the lowest number of terms, the solution having positive integral, minimum values is easily found :

$$a = 1, \qquad h = 1. \tag{22.6}$$

Equation (22.5, T), which has one additional term and where the value established for a is introduced, becomes :

$$-b + 2e = 1 \tag{22.7}$$

and admits the following solution with positive, integral, minimum values :

$$b = 1, \qquad e = 1. \tag{22.8}$$

Equations (22.5, L), which has the greatest number of terms, takes the form

$$c = 2 \tag{22.9}$$

and represents at the same time the value of the last unknown.

Relation (22.4), where the values found for the quantity exponents are introduced, takes the form

$$v = k\frac{d^2\, g\, \rho}{\eta}. \tag{22.10}$$

To check the result obtained, one should estimate the value of the numerical factor of proportionality k, which is done on the basis of experimental data ; now, the relation can take its final form and be used in practice. If the above relation is compared with that derived through the analytical method, we shall see that it is correct.

★

When deriving the physical relations for phenomena described by a greater number of measurable quantities, the use of mechanized means of calculation is recommended ; general dimensional analysis leads to an algorithm. In Appendix 4 a brief description of an algorithm to work with the general dimensional analysis is given.

22.3 A Comparison with Restricted Dimensional Analysis

For the sake of comparison, a few applications will be discussed; they will first be treated by general dimensional analysis and then by the restricted form of the same.

Case 1. A comparison is sought between the relation derived from physical quantities obtained by restricted dimensional analysis when $i^* = 1$, with the relation obtained by general dimensional analysis, on the basis of a system of units having the same number of fundamental quantities. For this purpose, let us consider the relation established for the velocity of fall of a sphere through a viscous medium (22.10); the substitution

$$\gamma = g\ \rho, \tag{22.11}$$

where γ is the apparent specific weight, will lead to the following function of $m^* = 4$ variables:

$$f(d, \gamma, \eta, v) = 0. \tag{22.12}$$

The dimensional equations of the variables will be written in the system of units with $d^* = 3$ fundamental quantities and result from the following dimensional matrix:

$$\begin{array}{c|cccc|} & d & \gamma & \eta & v \\ \hline L & 1 & -2 & -1 & 1 \\ M & 0 & 1 & 1 & 0 \\ T & 0 & -1 & -1 & -1 \end{array} \tag{22.13}$$

whose rank is $r^* = 3$.

The number of non-dimensional products of powers is $i^* = 1$.

The relationship between variables is written according to Rayleigh's method in the form

$$v = k\ d^a\ \gamma^b\ \eta^c. \tag{22.14}$$

The condition of dimensional homogeneity for relation (22.14) is expressed by the non-homogeneous system of linear equations :

$$\begin{aligned} a - 2b - c &= 1 \quad \text{(L)} \\ b + c &= 0 \quad \text{(M)} \\ -2b - c &= -1 \quad \text{(T)}. \end{aligned} \tag{22.15}$$

The solution of the determinate system is :

$$a = 2, \qquad b \ \text{[illegible]} \qquad c = 1. \tag{22.16}$$

Hence, the solution matrix is :

	a	b	c	
	d	γ	η	v
Π	-2	-1	1	1

(22.17)

The relation, as expressed by the physical quantities, is easily derived :

$$v = k \frac{d^2 \gamma}{\eta}. \tag{22.18}$$

According to general dimensional analysis, the relation between variables is written in the form :

$$\eta^a v^b = k\, d^c \gamma^e. \tag{22.19}$$

The condition of dimensional homogeneity for relation (22.19) is expressed by the homogeneous system of linear equations :

$$\begin{aligned} -a + b - c + 2e &= 0 \quad \text{(L)} \\ b - e &= 0 \quad \text{(M)} \\ -a - b + 2e &= 0 \quad \text{(T)} \end{aligned} \tag{22.20}$$

built up from $e^* = 3$ equations with $n^* = 4$ unknowns ; hence, this system of equations is an indeterminate one.

To solve the system of equations (22.20), let us consider equation (22.20, M) having the lowest number of terms; this equation admits the solution:

$$b = 1, \qquad e = 1. \tag{22.21}$$

Equation (22.20, T) having one additional term, leads to the solution

$$a = 1. \tag{22.22}$$

Equation (22.20, L), having the greatest number of terms, leads to the solution

$$c = 2. \tag{22.23}$$

Relation (22.19), where the above values are introduced, becomes

$$v = k \frac{d^2 \gamma}{\eta}. \tag{22.18}$$

As can be seen, the results are identical.

Case 2. A comparison is sought between relation (22.18), written for the physical quantities obtained by restricted dimensional analysis, and the relation obtained through general dimensional analysis on the basis of a system of units with fewer fundamental quantities.

The dimensional equations of the variables are written in the system of units with $d^* = 2$ fundamental units and result from the following dimensional matrix:

	d	γ	η	v
L	1	1	2	1
T	0	−4	−3	−1

(22.24)

The condition of dimensional homogeneity for relation (22.24) is expressed by the homogeneous system of linear equations:

$$\begin{aligned} 2a + b - c - r &= 0 \quad (\mathrm{L}) \\ -3a - b + 4e &= 0 \quad (\mathrm{T}) \end{aligned} \tag{22.25}$$

built up from $e^* = 2$ equations with $n^* = 4$ unknowns; this system of equations is an indeterminate one.

To solve the system of equations (22.25), use shall be made of equation (22.25, T) which has the lowest number of terms; this equation admits the solution:

$$a = 1, \qquad b = 1, \qquad c = 1. \tag{22.26}$$

Equation (22.25, L) leads to solution

$$c = 2. \tag{22.27}$$

As seen, the values of quantity exponents are the same as those from relation (22.18).

Case 3. A comparison is sought between that relation which in restricted dimensional analysis has $i^* = 2$ when $d^* = 3$ and $i^* = 1$ when $d^* = 4$, and the relation derived through general dimensional analysis, when $d^* = 2$.

Therefore, let us consider the same physical phenomenon, that is a sphere falling through a viscous medium, for which the function with $m^* = 4$ variables is used:

$$f(d, g, \rho, \eta, v) = 0. \tag{13.3}$$

The dimensional analysis calculus was described in Section III. 13.2. The relation, as expressed by the physical quantities and the non-dimensional product of powers, has the form

$$v = k \frac{\eta}{d \rho} \varphi\left(\frac{d^3 g \rho^2}{\eta^2}\right). \tag{13.15}$$

When written in the form

$$v = k \frac{\eta}{d \rho} \left(\frac{d^3 g \rho^2}{\eta^2}\right)^b, \tag{22.28}$$

this solution leads to the correct relation, if $b = 1$.

Relation (13.15) contains a function φ, which cannot be determined by dimensional analysis. To eliminate this indeterminate function, we shall adopt a system of units with $d^* = 4$ fundamental quantities. The dimensional analysis calculus will lead to the relation-solution expressed by physical quantities; it is of the form

$$v = k \frac{d^2 g \rho}{\eta}. \tag{22.10}$$

In order to apply general dimensional analysis, use is made of the dimensional equations of variables from function (1.33), written in the system of units with $d^* = 2$ fundamental quantities; they result from the following dimensional matrix:

$$\begin{array}{c|ccccc|} & d & g & \rho & \eta & v \\ \hline L & 1 & 1 & 0 & 2 & 1 \\ T & 0 & -2 & -2 & -3 & -1 \end{array} \tag{22.29}$$

The relation between variables is written in the form

$$\eta^a v^b = k d^c g^e \rho^h \tag{22.4}$$

where unknown quantity exponents were ascribed to all variables.

The condition of dimensional homogeneity of relation (22.4) is expressed through the homogeneous system of linear equations:

$$\begin{aligned} 2a + b - c - e &= 0 \quad (L) \\ -3a - b + 2e + 2h &= 0 \quad (T) \end{aligned} \tag{22.30}$$

containing $e^* = 2$ equations with $n^* = 5$ unknowns; hence, the system of equations (22.30) is an indeterminate one.

The system of equations (22.30) will be solved by applying the rule already mentioned in this work. In such a case, both equations will hold the same number of terms. Let us first consider, say, equation (22.30, L); the solutions obtained are

$$\begin{aligned} &a = 1, \quad b = 1, \quad c = 1, \quad e = 2, \\ &a = 1, \quad b = 1, \quad c = 2, \quad e = 1. \end{aligned} \tag{22.31}$$

But this result is ambiguous. Equation (22.30, T) leads to the solution :

$$a = 1, \qquad b = 1, \qquad e = 1, \qquad h = 1. \tag{22.32}$$

As a consequence, the second solution is correct (22.31), since the unknown e has the same value as solution (22.32). Had equation (22.30, T) been solved first, no ambiguity would have resulted.

The correct values of quantity exponents are now introduced in relation (22.4)

$$v = k \frac{v^2 g \rho}{\eta}, \tag{22.10}$$

i.e. an already known result.

Case 4. A comparison is sought between the relation of physical quantities when $i^* > 1$ (for which the values of the exponents of non-dimensional products of powers are determined by Huntley's method) and the relation derived by general dimensional analysis.

To this end, let us reconsider the following problem : a projectile is launched at a horizontal speed to a height h ; what is the distance to the point of fall? The function drawn up for dimensional analysis calculus contains $m^* = 5$ variables :

$$f(m, v, h, x, G) = 0. \tag{12.49}$$

Taking into account relation (12.2), we finally achieve the result

$$x = k\, h \left(\frac{hg}{v^2} \right)^d, \tag{12.56}$$

where the non-dimensional product of powers has a unknown exponent (Section III.12.3).

When deriving a physical relation between variables, the author used Huntley's method (Section III.12.3) which uses dimensional equations with oriented fundamental quantities. The dimensiona analysis computations carried out for the same function of variables and relation (12.2) led to the following relation

$$x = kv\sqrt{\frac{h}{g}}. \qquad (12.61)$$

In order to use the general dimensional analysis, the dimensional equations of variables are written, say, in the system of units with $d^* = 2$ fundamental quantities; they result from the following dimensional matrix:

	m	v	h	x	G
L	3	1	1	1	4
T	−2	−1	0	0	−4

(22.33)

The relation between variables is written in the form

$$x^a\, G^b = k\, m^c\, v^d\, h^e \qquad (22.34)$$

all variables having unknown quantity exponents.

The condition of dimensional homogeneity for relation (22.34) is expressed by the homogeneous system of linear equations:

$$\begin{aligned} a + 4b - 3c - d - e &= 0 \qquad (\mathrm{L}) \\ -4b + 2c + d &= 0 \qquad (\mathrm{T}) \end{aligned} \qquad (22.35)$$

built up with $e^* = 2$ equations with $n^* = 5$ unknowns; this system of equations is an indeterminate one.

In order to solve the system of equations (22.35), let us consider equation (22.35, T), with the lowest number of terms; the solution obtained is:

$$b = 1, \qquad c = 1, \qquad d = 2, \qquad (22.36)$$

having positive, integral, minimum values. Equation (22.35, L) becomes

$$a - e = 1 \qquad (22.37)$$

and admits the following solution having values of like nature:

$$a = 2, \qquad e = 1. \tag{22.38}$$

Relation (22.34) takes the form

$$x^2 = kv^2 \frac{mh}{G}. \tag{22.39}$$

Relations (12.2) and (2.39) lead to the final relation

$$x^2 = kv^2 \frac{h}{g}; \tag{22.40}$$

relation (12.61), derived through Huntley's method, implicates relation (22.40), derived through the general dimensional analysis.

General dimensional analysis still yields a correct result if another system of units is applied. Moreover, the correct result can also be reached by using the function of variables given in relation (12.61).

Case 5. A comparison is sought between the relation of physical quantities when $i^* > 1$, for which the values of the exponents of non-dimensional products of powers are determined by the algebraic method, with the relation derived by general dimensional analysis.

With this purpose in view, let us consider the physical phenomenon of bending a supported beam for which the implicit function of $m^* = 5$ variables is

$$f(l, I, E, P, s) = 0. \tag{16.15}$$

For a system of units with $d^* = 3$ fundamental quantities, the relation between $i^* = 3$ non-dimensional products of powers was derived by restricted dimensional analysis. The relation-solution expressed by the physical quantities and by the non-dimensional products of powers is of the form

$$s = k\, l \left(\frac{I}{l^4}\right)^b \left(\frac{l^2 E}{P}\right)^c. \tag{16.21}$$

Experimental data recorded helped to calculate the values of the numerical factor and of the exponents of non-dimensional products of powers, the results being :

$$k \approx 0.02, \qquad b \approx -1, \qquad c \approx -1. \tag{16.23}$$

The relation expressing the sag of the beam is of the form (Section III.16.2) :

$$s = 0.02 \frac{l^3 P}{IE}\cdot \tag{16.24}$$

In order to apply general dimensional analysis, let us consider the dimensional equations of variables in the function (16.15) written, for example, in the system of units with $d^* = 2$ fundamental quantities ; they result from the following dimensional matrix :

	l	I	E	P	s
L	1	4	2	4	1
T	0	0	−4	−4	0

(22.41)

The relation between variables is written in the form

$$s^a I^b E^c = k P^d l^e, \tag{22.42}$$

where unknown quantity exponents were ascribed to all variables.

The condition of dimensional homogeneity for relation (22.42) is expressed by the homogeneous system of linear equations :

$$\begin{aligned} a + 4b + 2c - 4d - e &= 0 \qquad \text{(L)} \\ -4c + 4d &= 0 \qquad \text{(M)} \end{aligned} \tag{22.43}$$

built up with $e^* = 2$ equations with $n^* = 5$ unknowns ; hence, the system of equations (22.43) is an indeterminate one.

In order to solve the system of equations (22.43), let us consider equation (22.43, M) with the lowest number of terms ; the solution in positive, integral, minimum values is :

$$c = 1, \qquad d = 1. \tag{22.44}$$

Equation (22.43, L) becomes

$$a + 4b - e = 2 \tag{22.45}$$

and admits the following solution in positive, integral, minimum values :

$$a = 1, \qquad b = 1, \qquad e = 3. \tag{22.46}$$

Relation (22.42), where the values determined for the quantity exponents are introduced, takes the form

$$s = k\frac{l^3 P}{IE}. \tag{22.47}$$

The value of the numerical factor of proportionality k may be found by experimental measurements.

A comparison with relation (16.24) will show that the result found through general dimensional analysis is correct. This result can be reached with other systems of units as well.

22.4 Comparison with the Analytical Method

In the author's opinion a parallel study conducted on a physical phenomenon using the analytical method and both the restricted and general dimensional analysis is of interest for the sake of the comparison. With this end in view, let us consider the physical phenomenon of vapour condensation.

The flow of heat during vapour condensation is associated with the change in the fluid state of aggregation and differs from that which takes place in the presence of one single phase. When the vapours are brought into contact with a wall with a temperature lower than their saturation temperature, the condensation phenomenon takes place. The resulting liquid is deposited on the wall, in the form of either a continuous film or droplets ; this process is called film condensation or dropwise condensation, respectively. The latter phenomenon is possible only when the condensate does not wet the heating (or cooling) surface ; when pure vapours condense from a wall-wetting liquid, a continuous film will always be obtained over a clean surface. As a general rule, dropwise condensation is unstable. In what follows, we shall consider the phenomenon of the flow of heat on film condensation.

According to *the analytical method*, the quantity of heat, q_x, transferred throughout the unit area is determined on the basis of Newton's law:

$$q_x = \alpha_x \Delta \theta, \tag{22.48}$$

where α_x is the heat-transfer coefficient, while $\Delta\theta$ is the temperature difference.

The heat released during the condensation process has to be conveyed towards the wall by the condensate film; inasmuch as the liquid flow is laminar, the heat transfer takes place through conduction. It is admitted that the temperature of condensate particles in the close vicinity of the wall is equal to temperature θ_w of the wall, while the condensate particles in contact with the vapours are at the condensation temperature θ_s. It is assumed that these temperatures are constant over the whole surface. The heat transferred through the unit area is estimated with Fourier's relation:

$$q_x = -\frac{\lambda}{\delta_x}\Delta\theta = -\frac{\lambda}{\delta_x}(\theta_s - \theta_w), \tag{22.49}$$

where λ is the condensate thermal conductivity, while δ_x is the liquid film thickness across section x.

The ratio of the transferred heat q_x to the latent heat of vaporization r gives the amount of condensate G_x, i.e.

$$G_x = \frac{q_x}{r} = -\frac{\lambda}{\delta_x}\cdot\frac{1}{r}(\theta_s - \theta_w). \tag{22.50}$$

Relations (22.48) and (22.49) lead to

$$\alpha_x = -\frac{\lambda}{\delta_x}, \tag{22.51}$$

which means that for the estimation of coefficient α_x, the condensate film thickness δ_x should be known.

Following Nusselt, the thickness of the condensate film is obtained from the flow equation, written in the form

$$\frac{d^2 v_x}{dy^2} = -\frac{\gamma}{\eta}, \tag{22.52}$$

where the symbols represent the vertical velocity of flow of the condensate v_x, the specific weight γ, and the dynamic viscosity η; x and y represent the two cartesian co-ordinate axes, x being in a vertical plane, while y is normal to this plane.

For the amount of condensate G_x flowing in the unit time through section x of a plate having a unit width, one can also write the relation :

$$G_x = v_{x,m}\, \gamma\, \delta_x = \frac{\gamma^2\, \delta_x^3}{3\eta}, \tag{22.53}$$

where $v_{x,m}$ is the liquid mean velocity.

On the basis of both the above equations and after some cumbersome mathematical analysis calculations, we shall find the relation for the mean heat-transfer coeffcient α :

$$\alpha = \frac{4}{3} \sqrt[4]{\frac{r\, \gamma^2\, \lambda^3}{4h\, \eta\, \Delta\theta}}, \tag{22.54}$$

where h is the wall height.

Relation (22.54) leads to values higher by 20 to 22 % than those found by experimental measurements [3.11][22.1].

In *restricted dimensional analysis*, this phenomenon is studied with the help of the heat transfer equation (22.51), the flow equation (22.52), and the conduction equation :

$$v\, \frac{\partial \theta}{\partial x} = a\, \frac{\partial^2 \theta}{\partial y^2}, \tag{22.55}$$

where a is the thermal diffusivity, and of the thermal balance equation :

$$r\, \gamma\, v\, \delta = \alpha\, h\, \Delta\, \theta = \frac{\lambda}{\delta}\, h\, \Delta\theta, \tag{22.56}$$

22.1 p. 118.

where due consideration is paid to the change of state, between the vapour and the liquid phases at the limit.

From the equations above, the following function of variables results :

$$f(\alpha, \Delta\theta, \lambda, l, v, \gamma, \eta, r) = 0, \tag{22.57}$$

where l represents a characteristic length. Now, restricted dimensional analysis will readily lead to the relation

$$Nu = k\,\varphi\,(K, Pr, Ga), \tag{22.58}$$

where symbols stand for Nusselt, K (no name), Prandtl and Galilei invariants, respectively.

Relation (22.58) is a rather indefinite account of the phenomenon, since restricted dimensional analysis cannot assess the values of the numerical factor and of the three exponents of non-dimensional products of powers. In such a situation, one should

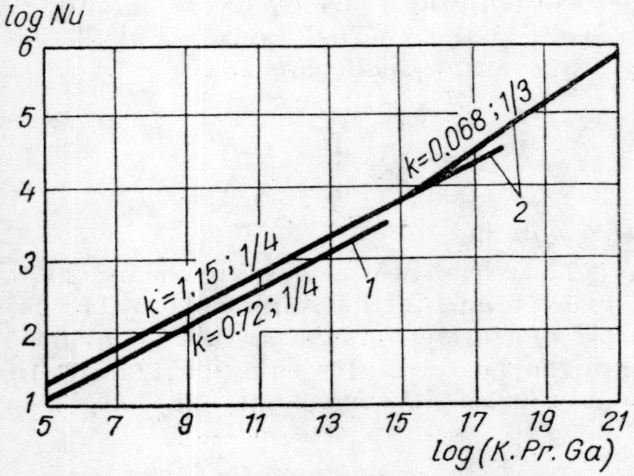

Fig. 22.1 Vapour condensation in (1) horizontal and (2) vertical pipes

have recourse to experimental data ; these are illustrated in Fig. 22.1. Using the diagram from Fig. 22.1, relation (22.58) becomes, i. e :

$$Nu = k\,(K \cdot Pr \cdot Ga)^{1/4} \tag{22.59}$$

and the numerical factor of proportionality will have the values $k = 0.72$ for horizontal, and $k = 1.15$ for vertical tubes. As can easily be seen, relation (22.54), derived by the analytical method, is similar to relation (22.59). With vertical tubes and for values of the product $K \cdot Pr \cdot Ga > 10^{15}$, relation (22.58) takes the form

$$Nu = k\,(K \cdot Pr \cdot Ga)^{1/3}. \tag{22.60}$$

Now, in *general dimensional analysis*, this physical phenomenon is studied by means of variables:

$$f\,(\alpha, \Delta\vartheta, h, \eta, \lambda, \gamma, r) = 0. \tag{22.61}$$

The relation between variables is written in the form

$$\alpha^b\,\Delta\theta^c\,h^d\,\eta^e = k\lambda^i\,\gamma^j\,r^l. \tag{22.62}$$

The condition of dimensional homogeneity for relation (22.62) is expressed by the indeterminate, homogeneous system of linear equations:

$$\begin{aligned}
-2b + d - 2e + i + 3j &= 0 \qquad (L)\\
-b + e + i &= 0 \qquad (T)\\
e - j + l &= 0 \qquad (F)\\
-b + c + i &= 0 \qquad (\Theta)\\
b - i - l &= 0 \qquad (Q).
\end{aligned} \tag{22.63}$$

As can be seen, four of the equations have three terms each and, of these, three equations differ only by one unknown. From equations (22.63. T), (22.63, Θ) and (22.63, Q) we obtain:

$$b = e + i = c + i = l + i \tag{22.64}$$

and the following solution results:

$$b = 2, \qquad c = 1, \qquad e = 1, \qquad i = 1, \qquad l = 1, \tag{22.65}$$

having positive, integral, minimum values.

Equation (22.63, F), having the same number of terms, leads to the solution

$$j = 2. \tag{22.66}$$

Equation (22.63, L) leads to a negative d, which contradicts relation (22.62).

On the basis of equations (22.63), we admit the solution with positive, integral, minimum values :

$$c = 1, \qquad e = 1, \qquad l = 1. \tag{22.67}$$

Equation (22.63, F) leads to the solution

$$j = 2 \tag{22.68}$$

which is of like nature. Equation (22.63, L) takes the form

$$d + i + 4 = 2b. \tag{22.69}$$

Equation (22.63, T) and equation (22.69) lead to equation

$$d + 2 = i, \tag{22.70}$$

which admits the solution :

$$d = 1, \qquad\qquad i = 3, \tag{22.71}$$

having positive, integral, minimum values. Any of equations (22.63,T, Θ, Q) leads to a final solution which is similar :

$$b = 4. \tag{22.72}$$

Relation (22.62) becomes

$$\alpha^4 = k \frac{\lambda^3 \gamma^2 r}{\Delta\theta\, h\, \eta}\,. \tag{22.73}$$

A comparison with relation (22.54) will show that relation (22.73) is correct.

22.5 A Few Particular Cases

The functions describing physical phenomena include various dimensional variables : physical quantities, universal constants and characteristic constants. Each variable has its own dimensional equation. The frequency of fundamental quantities is an aleatory quantity and leads to truncated triangular systems of equations between dimensional and quantity exponents. On account of the very wide variety of situations, which may arise in practice, no rules can be stated in this respect. However, it was actually found that for a given physical phenomenon, the condition of dimensional homogeneity for the corresponding rational relation might be expressed by a system of equations containing several equations e_k^*, with a constant number of terms, i.e.

$$e_k^* > 1. \tag{22.74}$$

When solving such a system of equations, it is obvious that any of the equations having the same number of terms can be solved before the others of the same kind. As a consequence, due consideration should be paid to all such equations, proceeding as follows :

— each group of equations having the same number of terms should be solved several times by changing the order of resolution ;

— substitutions in these equations should be operated so as to obtain only one equation.

It was found in practice that in some particular cases the correct solution for quantity exponents having positive, integral, minimum values is obtained after 2 to 4 trials. The correct solution is found by checking the result yield by dimensional analysis against the actual measurements.

In the case of functions of several variables of like nature, the result of dimensional analysis can sometimes be ambiguous ; indeed, 2 to 4 solutions can be obtained for the quantity exponents having positive, integral minimum values. The correct relation is again chosen on the basis of experimental data.

Practice has showed that in the case of approximate functions 2 to 4 solutions for the quantity exponents having positive, integral, small values should be considered to afford a comparison between the results of dimensional analysis and those supplied by actual measurements; each relation corresponds to a more limited range and hence, 2 to 4 relations are needed to cover the whole range of actual values. In doing so, a more accurate correlation is sought between the values reached by general dimensional analysis and the measured values.

As given, systems of units with several fundamental quantities are available. Thus, for instance, in mechanics: $d^* = 2$, $d^* = 3$ and $d^* = 4$. In dimensional analysis where, generally, the systems of equations between dimensional and quantity exponents are indeterminate, the systems of units with a higher number of fundamental quantities are recommended. Thus, an increased efficiency of the dimensional analysis is achieved, since the number of equations expressing the condition of dimensional homogeneity is higher.

If a given physical phenomenon is to be studied by dimensional analysis, the study should be carried out for various systems of units. These should be chosen so as to avoid errors; if, for instance, in mechanics $d^* = 2$ is used, then a dimensionally homogeneous relation might result even though the list of parameters contains only one physical quantity specific to the field; obviously, such a result is incorrect.

22.6 On the Importance of the Numerical Factor

Whenever a physical phenomenon is studied through general dimensional analysis, the value of the numerical factor should be necessarily determined, since a dimensionally homogeneous relation is physically correct if, and only if, the value of this factor is a constant. When the numerical factor has a definite value, the relation is complete and therefore can be used in practice.

At the same time, the numerical factor can be used to check the elements and results obtained using the method of dimensional analysis, such as:

— if the function of variables contains the physical quantities involved in the physical phenomenon under study;

— if the preliminary distribution of quantities among the members of the monomial relation is correct;

— if the system of units used is adequate;

— if the solution yielding integral, minimum, positive values for the quantity exponents is checked by the experimental results;

— if the solution containing integral, small, positive values adopted for the quantity exponents checks well with experimental data.

23 Retrieved Physical Relations

23.1 Elongation under Centrifugal Stresses

Let us consider a ring of radius r in free rotation at a velocity v. The ring is made out of a material of density ρ and a modulus of elasticity E. Owing to the centrifugal force, the ring radius increases by Δr. The elongation varies as a direct function of radius r, velocity v and density ρ; and as a reverse function of the modulus of elasticity E [5.35].

The analytical method leads to the relation

$$\Delta r = \frac{r\,v^2 \rho}{E}. \tag{23.1}$$

The restricted dimensional analysis leads to, say, the relation

$$\Delta r = kr\,\varphi\left(\frac{E}{v^2 \rho}\right), \tag{23.2}$$

where an indeterminate function intervenes.

The general dimensional analysis is applied to the physical model of the phenomenon described above. The implicit function of $m^* = 5$ variables is

$$f(r,\ v,\ \rho,\ E,\ \Delta r) = 0. \tag{23.3}$$

The relation having the variables distributed between the two members is

$$f_1(E,\ \Delta r) = f_2(r,\ v,\ \rho). \tag{23.4}$$

The dimensional matrix of variables for the $d^* = 3$ fundamental quantities is:

$$\begin{array}{c|ccccc|} & r & v & \rho & E & \Delta r \\ \hline \mathrm{L} & 1 & 1 & -3 & -1 & 1 \\ \mathrm{M} & 0 & 0 & 1 & 1 & 0 \\ \mathrm{T} & 0 & -1 & 0 & -2 & 0 \end{array} \tag{23.5}$$

The frequency of fundamental quantities in the dimensional equations of variables are:

$$\nu_L^* = 0.55,\ \ \nu_M^* = 0.22,\ \ \nu_T^* = 0.22. \tag{23.6}$$

Solution 1 uses the repeated explicit form with various variables to find the values of the quantity exponents.

Let us consider, for instance, the fundamental quantity M, having a minimum frequency. Relation (23.4) is written in the form

$$E^a f(\Delta r) = k\rho^b f(r,\ v). \tag{23.7}$$

Introducing the dimensional equations of variables E and ρ in relation (23.7) results in:

$$(\mathrm{L}^{-1}\mathrm{MT}^{-2})^a f(\Delta r) = k(\mathrm{L}^{-3}\mathrm{M})^b f(r,\ v). \tag{23.8}$$

The condition of dimensional homogeneity for relation (23.8) with respect to the fundamental quantity M is expressed by the equation

$$a = b. \tag{23.9}$$

Variable E is put in an explicit form and equation (23.9) is solved to obtain the solution:

$$a = 1, \quad b = 1. \tag{23.10}$$

Let us now consider the fundamental quantity T which, in this case, has the same frequency. Based on the values found above, relation (23.7) can be written as

$$E f(\Delta r) = k\ \rho v^c f(r). \tag{23.11}$$

Introducing the dimensional equations of variables E, ρ and v in relation (23.11) results in:

$$\mathrm{L}^{-1}\,\mathrm{MT}^{-2} f(\Delta r) = k\,\mathrm{L}^{-3}\mathrm{M}\,(\mathrm{LT}^{-1})^c f(r). \tag{23.12}$$

The condition of dimensional homogeneity for relation (23.12) with respect to the fundamental quantity T leads to

$$c = 2. \tag{23.13}$$

Let us consider the fundamental quantity L, which has the highest frequency. On the basis of the results above, relation (23.11) can be written in the form

$$E\Delta r^d = k\rho\, v^2\, r^e. \tag{23.14}$$

Introducing the dimensional equations of variables in relation (23.14) results in

$$\mathrm{L}^{-1}\mathrm{MT}^{-2}\cdot\mathrm{L}^d = k\,\mathrm{L}^{-3}\,\mathrm{M}\cdot\mathrm{L}^2\,\mathrm{T}^{-2}\cdot\mathrm{L}^e. \tag{23.15}$$

The condition of dimensional homogeneity for relation (23.15) with respect to the fundamental quantity L is expressed by the equation

$$d = e. \tag{23.16}$$

Variable Δr is put in an explicit form and equation (23.16) is solved, the result being:

$$d = 1, \qquad e = 1. \tag{23.17}$$

Introducing in relation (23.4) the values found for all quantity exponents results in

$$\Delta r = k\frac{rv^2\rho}{E}. \tag{23.18}$$

The same result can be reached when the general dimensional analysis calculus starts with the fundamental quantity T, which has the same frequency as the fundamental quantity L.

Solution 2 uses the general procedure for establishing the values of the quantity exponents.

The distributed relation of $m^* = 5$ variables is (23.4).

The dimensional matrix of variables for $d^* = 3$ fundamental quantities is (23.5).

The monomial relation having unknown exponents for all variables is

$$E^a \Delta r^b = k\, r^c v^d \rho^e. \tag{23.19}$$

The system of linear equations expressing the condition of dimensional homogeneity is:

$$\begin{aligned} -a + b - c - d + 3e &= 0 \quad \text{(L)} \\ a - e &= 0 \quad \text{(M)} \\ -2a + d &= 0 \quad \text{(T)}. \end{aligned} \tag{23.20}$$

In compliance with the rule (I) governing the resolution of Diophantine systems of equations, the solution of the indeterminate system of equations (23.20) is:

$$\begin{aligned} &a = 1, \qquad e = 1; \\ &d = 2; \\ &b = 1, \qquad c = 1. \end{aligned} \tag{23.21}$$

The relation, as expressed by the physical quantities,

$$\Delta r = k \frac{r v^2 \rho}{E} \tag{23.18}$$

is as above.

Solution 3 makes use of an equivalent accurate function and of the same general procedure for finding the values of the quantity exponents. In the function of variables (23.3) the density ρ is substituted by the specific weight γ and by the normal acceleration of free fall g.

The distributed relation of $m^* = 6$ variables is

$$f_1(E, \Delta r, g) = f_2(r, v, \gamma). \tag{23.22}$$

The dimensional matrix of variables for $d^* = 3$ fundamental quantities is:

	r	v	γ	g	E	Δr
L	1	1	−2	1	−1	1
M	0	0	1	0	1	0
T	0	−1	−2	−2	−2	0

(23.23)

The monomial relation having unknown exponents for all variables is:

$$E^a \, \Delta r^b \, g^c = k \, r^d \, v^e \, \gamma^h. \tag{23.24}$$

The system of linear equations expressing the condition of dimensional homogeneity is:

$$\begin{aligned} -a + b + c - d - e + 2h &= 0 \quad \text{(L)} \\ a - h &= 0 \quad \text{(M)} \\ -2a - 2c + e + 2h &= 0 \quad \text{(T).} \end{aligned} \tag{23.25}$$

The solution for the indeterminate system of equations, according to the rule (I) set out for the resolution of Diophantine systems of equations is:

$$\begin{aligned} a &= 1, \qquad & h &= 1; \\ c &= 1, \qquad & e &= 2; \\ b &= 1, \qquad & d &= 1. \end{aligned} \tag{23.26}$$

The relation, as expressed by the physical quantities, is:

$$\Delta r = k \frac{r \, v^2 \gamma}{E g}. \tag{23.27}$$

It can be readily seen that relation (23.18) is equivalent to relation (23.27), on the basis of the known relation:

$$\gamma = g \, \rho. \tag{22.11}$$

The value of the numerical factor of proportionality k in relations (23.18) and (23.27) is found by experimental measurements.

The result found by the theoretical method can be obtained through the restricted dimensional analysis only if for the exponent of the non-dimenisonal product of powers from relation (23.2) a negative unit value is obtained, through any method. The correct result was obtained through general dimensional analysis as well.

23.2 Turbulence — Free Convection across a Plate within an Unlimited Space

Let us consider a plate of length l. A fluid film flows at velocity v along this plate. The fluid has a kinematic viscosity ν, a heat conductivity λ and a thermal diffusivity a.

The normal acceleration of free fall is g. In this situation, a heat exchange through free convection takes place. What is the relation for the coefficient of heat flow?

The analytical method has failed as yet to supply a relation to describe this physical phenomenon.

The restricted dimensional analysis leads, for example, to the relation:

$$Nu = \varphi(Re,\ Ga,\ Pr) \tag{23.28}$$

between the non-dimensional products of powers of Nusselt, Reynolds, Galilei and Prandtl. The form of function φ cannot be determined with this method.

The general dimensional analysis is based on the physical model described above. The implicit function of $m^* = 7$ variables is

$$f(l,\ v,\ \nu,\ \lambda,\ a,\ g,\ \alpha) = 0. \tag{23.29}$$

The distributed relation of $m^* = 7$ variables is

$$f_1(\nu,\ a,\ \alpha) = f_2(l,\ v,\ \lambda,\ g). \tag{23.30}$$

The monomial relation having unknown exponents for all variables is:

$$\nu^b\ a^c\ \alpha^d = kl^e\ v^h\ \lambda^i\ g^j. \tag{23.31}$$

The solution for the indeterminate system of equations expressing the condition of dimensional homogeneity of the relation (23.31) is found by applying the rule (II) for solving Diophantine systems of equations. Only those solutions yielding positive, integral, small values are retained.

The solution matrix is:

	b ν	c a	d α	e l	h v	i λ	j g
R_1	2	1	1	3	1	1	1
R_2	1	2	1	3	1	1	1
R_3	2	1	2	2	1	2	1
R_4	1	2	2	2	1	2	1
R_5	2	1	3	1	1	3	1
R_6	1	2	3	1	1	3	1

(23.32)

The solution, as expressed by the physical quantities is written for each line of values of the solution matrix (23.32), that is R_1 R_2, ...

The value of the numerical factor of proportionality k will be estimated for the six relations on the basis of experimental results. The correct relation will be that where the value of the numerical factor is kept constant.

The relation found by applying the restricted dimensional analysis should be completed – through other procedures – with the values of the numerical factor and of the exponents of non-dimensional products of powers; the final relation is of the form [5.32]:

$$Nu = 0.01(Re.\ Ga.\ Pr)^{1/3}. \qquad (23.33)$$

This is R_5 found through general dimensional analysis.

23.3 Drop of Linear Pressure in Laminar Flows

Let us consider the permanent flow of a liquid within a restricted closed space. The liquid has a density ρ, a dynamic viscosity η and flows at a velocity v through a pipe of a diameter d. Across two sections of the fluid stream, placed at a distance l from one another, a pressure drop Δp is produced. The physical system described is actually sited at a place on the ground where g is the acceleration of the free fall.

The analytical method was applied by Hagen-Poiseuille to the study of this phenomenon ; by accurately solving the Navier-Stokes equations for the motion of real fluids, the following relation was derived :

$$\Delta p = 32\, v\, \eta \frac{1}{d^2}\,. \tag{23.34}$$

The experimental method has been used intensively. With experimental data supplied by Nikuradze it was possible to derive the dependence of the friction coefficient on the Reynolds number :

$$\lambda = f(v\, d\, \tau\, \eta^{-1}). \tag{23.35}$$

as shown in Fig. 23.1. Based on this dependence and on measurements conducted with water at a temperature of 0°C the data presented in Table 21.3 were found.

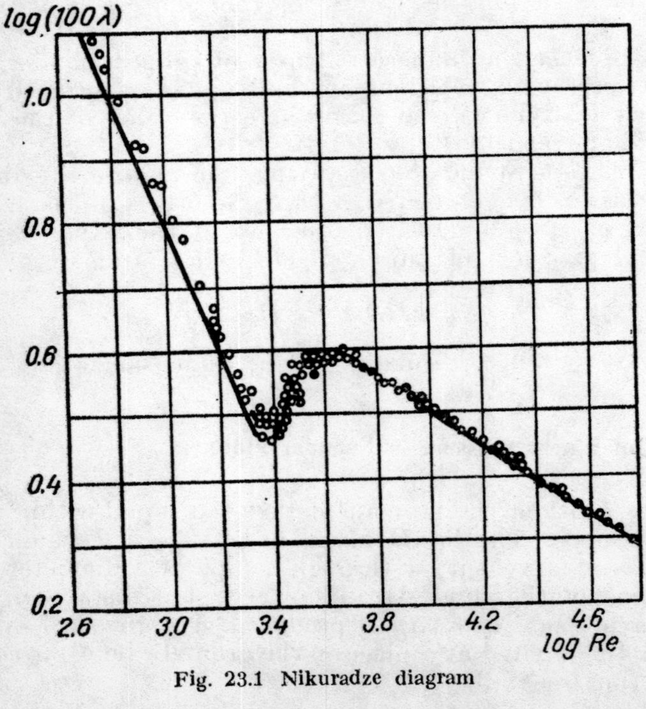

Fig. 23.1 Nikuradze diagram

TABLE 23.1. **Measured values for pressure loss**
$\rho = 102$ kgf·s²/m⁴
$\eta = 102.5\ 10^{-6}$ kgf.s/m²

v, m/s	d,m	l,m	p,kgf/m²
0.0219	0.05	15	0.806
0.0237	0.06	24	0.954
0.0268	0.07	30	0.990
0.0338	0.08	40	1.276
0.0385	0.09	50	1.440

Restricted dimensional analysis leads, for instance, to the relation

$$\Delta p = \lambda k\, v^2\, \rho\, \varphi\left(\frac{l}{d}\right), \tag{23.36}$$

where the friction coefficient λ — which is a non-dimensional quantity — as well as an indeterminate function φ are also included. As given according to Hagen Poiseuille:

$$\lambda Re = 64, \tag{23.37}$$

where Re is the Reynolds number. It can readily be seen that relations (23.36) and (23.37) lead to relation (23.34).

General dimensional analysis takes into consideration the function of variables used to describe the drop of linear pressure in the case of a laminar flow:

$$f(v, \rho, \eta, d, l, \Delta p, g) = 0. \tag{23.38}$$

But instead of density ρ, we can consider the specific weight γ, that is

$$f(v, \gamma, \eta, d, l, \Delta p, g) = 0, \tag{23.39}$$

the two quantities being correlated through the known relation $\gamma = g\rho$.

The problem to be solved is the form of either relation (23.38) or (23.39). From investigations carried out by the author, the conclusion was reached that relation (23.38) will not lead to the result looked for, i.e. a physical relation which could handle the results of the measurements. The correct result can be derived only from function (23.39).

The relation with $m^* = 7$ variables distributed among the two members and having unknown exponents for all variables is :

$$\left(\frac{\Delta p}{l}\right)^a d^b g^c = K v^e \gamma^h \eta^i ; \tag{23.40}$$

this relation paying due consideration to the fact that the pressure drop Δp varies linearly with the length l of the pipe.

The dimensional matrix of variables for $d^* = 3$ fundamental quantities is :

	Δp	d	g	v	γ	η	l
L	-1	1	1	1	-2	-1	1
M	1	0	0	0	1	1	0
T	-2	0	-2	-1	-2	-1	0

(23.41)

The system of $e^* = 3$ linear equations with $n^* = 6$ unknowns, expressing the condition of dimensional homogeneity for relation (23.40), is :

$$\begin{aligned} -2a + b + c &= e - 2h - i & \text{(L)} \\ a &= h + i & \text{(M)} \\ -2a - 2c &= -e - 2h - i & \text{(T)}. \end{aligned} \tag{23.42}$$

The solution of the indeterminate system of equations (23.42), according to the rule (I) for the resolution of Diophantine systems of equations, is :

$$a = 2, \quad b = 3, \quad c = 1, \quad e = 3, \quad h = i = 1. \tag{23.43}$$

The relation, as expressed through the physical quantities, is

$$\Delta p = K \left(\frac{v^3 \rho \eta l^2}{d^3}\right)^{0.5} . \tag{23.44}$$

The values of the numerical factor K are estimated from Table 23.1, the results being presented in Table 23.2. It will be seen that values of K vary within wide boundaries, namely up to 100 per cent; according to general dimensional analysis, this result will not be accepted and new elements have to be introduced.

Given that the criterion for truth is practice, the criterion for the theory — such as the inference above, made on a dimensional basis — is experiment. As a consequence, we shall take into consideration the experimental results reported by Nikuradze and observe that the value of factor K changes as a function of the Reynolds number: in its most general form this is written

TABLE 23.2. **Values of numerical factor K**

$\left(\frac{v^3\rho\eta l^2}{d^3}\right)^{0.5}$	Δp	K
0.594	0.806	1.460
0.803	0.954	1.186
0.970	0.990	1.021
1.500	1.276	0.850
1.916	1.440	0.753

$$K = k\,(Re)^x. \tag{23.45}$$

Relations (23.44) and (32.45) lead to relation:

$$\Delta p = k(Re)^x \left(\frac{v^3 \rho \eta l^2}{d^3}\right)^{0.5}. \tag{23.46}$$

Using the data from Tables 23.1 and 23.2, the following equations of the same type as (23.46) will be written:

$$\begin{aligned} 0.806 &= 0.594\,k \quad 602^x \quad (1)\\ 0.954 &= 0.803\,k \quad 794^x \quad (2)\\ 0.990 &= 0.970\,k\,1.053^x \quad (3)\\ 1.276 &= 1.500\,k\,1.515^x \quad (4)\\ 1.440 &= 1.916\,k\,1.930^x \quad (5) \end{aligned} \tag{23.47}$$

where the unknowns k and x will be found. After trying various combinations of two equations, the solutions are determined. The results of the calculations are reported in Table 23.3.

Now, the values of k and x vary within rather restricted boundaries; as a consequence, we can accept the following mean values: $k_m = 32.02$ and $x_m = -0.503$. For all practical purposes we shall consider:

$$k = 32, \quad x = -0.5. \quad (23.48)$$

TABLE 23.3. **Values of numerical factor k**

Equation	k	x
23.47. 1,2	31.6	−0.492
23.47. 2,3	31.0	−0.506
23.47. 3,4	33.2	−0.501
23.47. 4,5	34.3	−0.511
23.47. 3,5	30.0	−0.505

Relation (23.46) takes the final form:

$$\Delta p = \frac{32}{Re^{0.5}} \left(\frac{v^3 \rho \eta l^2}{d^3} \right)^{0.5} \cdot \quad (23.49)$$

Some simple calculations will show that relation (23.49) is of the same form as relation (23.34), which is already known.

24 Established Physical Relations

24.1 Magnetism of a Rotational Body

In 1947, Blackett suggested that terrestrial magnetism would be the result of its rotating motion. According to this assertion, any solid, rotating body is a magnet[24.1]. The variables helping to describe this phenomenon are: body radius r; body mass m; the angular velocity ω; the velocity of light c; and the gravitational constant k'. The magnetic field is denoted by H.

The analytical method failed to establish a relation between these variables. Blackett suggests for the expression of the magnetic field the empirical relation

$$H = k \frac{m\omega}{rc} \sqrt{k'} \cdot \quad (24.1)$$

To explain this relation, let us consider the magnetic moment of the rotating mass P, which is expessed by the relation

$$P = r^3 H \quad (24.2)$$

[24.1] Blackett, P. M. S., The magnetic field of massive rotating bodies, *Nature*, CLIX, May 1947.

and the moment of impulse U which, for a sphere, is expressed by the relation

$$U = \frac{2}{5} r^2 m\omega. \tag{24.3}$$

On the basis of relations (24.2) and (24.3), relation (24.1) takes the form

$$P = k \frac{U}{c} \sqrt{k'} \cdot \tag{24.4}$$

Relation (24.1) checks well for the Earth, the Sun and the Virgo 78 star. It is a particularly important relation for it is the first in physics to correlate the gravitational constant k' with the magnetic energy [5.3; 5.7].

Restricted dimensional analysis, on the basis of variables from relation (24.1),

$$f(r, m, \omega, c, k', H) = 0, \tag{24.5}$$

cannot lead to a relation between the variables; it yields instead a relation between non-dimensional products of powers since the number of variables is comparatively high. However on the basis of variables from relation (24.4);

$$f(c, k', P, U) = 0, \tag{24.6}$$

this method could yield a relation between the variables. The homogeneity of relations (24.1) and (24.4) can be easily seen.

The general dimensional analysis is applied to this physical phenomenon because the known relation is an empirical one. Let us consider the function of variables (24.5). According to Blackett, the magnetic field of the rotating body varies directly with its mass m, its angular velocity ω and with the gravitational constant k' and inversely with the body radius r and the velocity of light c.

The distributed relation of $m^* = 6$ variables is

$$f_1(r, c, H) = f_2(m, \omega, k'). \tag{24.7}$$

The dimensional matrix of variables for $d^* = 3$ fundamental quantities is

	r	m	ω	c	k'	H
L	1	0	0	1	3	−1/2
M	0	1	0	0	−1	1/2
T	0	0	−1	−1	−2	−1

(24.8)

and leads to three relations between the physical quantities, which means an ambiguous result.

The dimensional matrix of variables for $d^* = 2$ fundamental quantities is:

$$\begin{array}{c|cccccc|} & r & m & \omega & c & k' & H \\ \hline \mathrm{L} & 1 & 3 & 0 & 1 & 0 & 1 \\ \mathrm{T} & 0 & -2 & -1 & -1 & 0 & -2 \end{array} \tag{24.9}$$

The monomial relation having unknown exponents for all variables is

$$r^a c^b H^d = k m^e \omega^g k'^h. \tag{24.10}$$

The system of linear equations expressing the condition of dimensional homogeneity is:

$$\begin{aligned} a + b \ + d \ - 3e &= 0 \quad (\mathrm{L}) \\ -b - 2d + 2e + \ g &= 0 \quad (\mathrm{T}). \end{aligned} \tag{24.11}$$

The solution for the indeterminate system of equations, according to the rule (I) for solving Diophantine systems of equations is:

$$a = 1, \ b = 1, \ d = 1, \ e = 1, \ g = 1. \tag{24.12}$$

The relation as expresed by the physical quantities is

$$H = k \frac{m\omega}{rc} k'^h. \tag{24.13}$$

The dimensional equations of variables (24.8) are introduced in relation (24.13). Now, the condition for the dimensional homogeneity of any of the fundamental quantities L, M or T leads to the solution:

$$h = \frac{1}{2}. \tag{24.14}$$

And relations (24.13) and (24.14) lead to the final relation

$$H = k \frac{m\omega}{rc} \sqrt{k'}, \tag{24.1}$$

which is identical with that suggested by Blackett [5.10].

24.2 Diameter of Atomized Liquid Drops

The process of stationary combustion must ensure complete fuel combustion with a small excess of air. The condition is dependent also on the quality of the mixture, i.e. the homogeneity achieved between fuel and the combustion air. For liquid fuels, the homogenization quality is conditioned by the fineness of the atomizing process. In an actual plant, the fluid to be atomized is directed through an injector to a nozzle, whereby it is sprayed into the gaseous medium already injected into the combustion chamber; at this point, the liquid breaks up into fine individual particles which tend towards a certain stable form and change into spherical drops with a diameter d.

The analytic method takes into account the forces intervening in this physical process of liquid flow; therefore the differential equation of Navier-Stokes

$$\frac{\partial \vec{v}}{\partial t} + (\vec{v}\ \mathrm{grad})\vec{v} = g - \frac{1}{\rho}\ \mathrm{grad}\ p + \frac{\eta}{\rho}\Delta\vec{v} \qquad (24.15)$$

as well as the continuity equation:

$$\mathrm{div}\ \vec{v} = 0 \qquad (24.16)$$

are considered; here the symbols represent the speed v of the liquid, the normal acceleration of the free fall g, the pressure p, the liquid dynamic viscosity η, the time t and the liquid density ρ respectively. The liquid jet break-up is also favoured by the surrounding gas whose action on the liquid surface is manifest through normal pressures and tangential stresses. These forces may definitely affect the liquid jet break-up and hence the whole atomizing process. As a consequence, any analytical approach to the phenomenon of the surrounding gas flow should also start from both the Navier-Stokes differential equation and the continuity equation.

The analytic solution for the atomizing phenomenon implies the integration of the four-equation system, with several limit conditions. Among these, one is the Laplace equation which should be satisfied by each point of the liquid surface, and by the spherical drops as well; this equation is written in the form

$$p_l = p_g + \frac{4\sigma}{d}, \qquad (24.17)$$

where the new symbols l and g refer to the liquid and the surrounding gas respectively, while the unit stress has been noted by σ. Other limit conditions are the distributions of both the liquid and the gas velocities at the nozzle exit. Based on the investigations carried out by Rayleigh on an ideal liquid jet stability, on Taylor's experimental work on low-viscosity liquids, and on Haenlein's experimental work, which showed the effect of the jet speed and of the liquid physical properties on the final shape of the broken liquid jet, Weber derived the following equation for the jet break-up:

$$\mu^2 = \beta^2(1 - \beta^2)\frac{\sigma}{2\rho_l r} + \beta^3 f(\beta)\frac{v^2 \rho_g}{2\rho_l r^2} \tag{24.18}$$

where

$$\mu = \sqrt{\frac{\sigma}{6\rho_l r^3}} \tag{24.19}$$

and represents the rate of the excitation growth with time, and

$$\beta = \frac{2\pi r}{\lambda}, \tag{24.20}$$

where the new symbols represent: r — nozzle port radius; λ — wave-length for which the increase in surface disturbance has the highest rate. The function

$$f(\beta) = -i\,\frac{H_0^1(i\,\beta)}{H_1^1(i\,\beta)} \tag{24.21}$$

where H_0^1 and H_1^1 are Hankel's function according to Jahnke-Emde. Equation (24.19) includes several simplifying assumptions; actually, it fails to yield a practical solution to the problem of the diameter of atomized liquid drops, because wavelength λ is not accurately known. Accordingly, in the technical literature it is stated that a complete analytical solution of the atomizing problem is not possible although several theoretical investigations of this phenomenon have been carried out in the last 100 years [5.6; 5.32].

The experimental method was applied in order to reach a result which could be satisfactory for all practical purposes. For water atomized in air, experimental measurements lead to the characteristic curve shown in Fig. 24.1, which represents the relation $d = f(v)$.

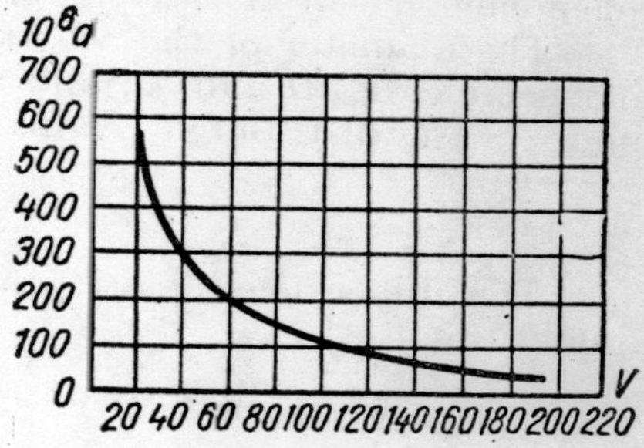

Fig. 24.1 Change in the mean diameter of atomized liquid particles

The restricted dimensional analysis considers the following function of $m^* = 7$ variables:

$$f(d, \eta_l, D, \sigma, \rho_g, v, \rho_l) = 0. \tag{24.22}$$

The computations lead to a relation between non-dimensional products of powers. The values of the numerical factor of the exponents of non-dimensional product of powers have been established by experimental measurements.

In the technical literature, the two relations recommended for the mean diameter of the atomized liquid drops are:

$$d = \left[1.24 + 0.0128\left(\ln\frac{\eta_l^2}{\rho_l \sigma D} + 12\right)\right] D \left(\frac{\sigma}{\rho_g v^2 D}\right)^{0.52} \tag{24.23}$$

for the range

$$\frac{\eta_l^2}{\rho_l \sigma D} < 0.0333 \tag{24.24}$$

and

$$d = \left(2.25 + 0.24 \ln \frac{\eta_l^2}{\rho_l \sigma D}\right) D \left(\frac{\sigma}{\rho_g v^2 D}\right)^{0.52} \tag{24.25}$$

for the range

$$\frac{\eta_l^2}{\rho_l \sigma D} > 0.0333. \tag{24.26}$$

In these relations, the new symbols represent the relative speed v of the liquid reported to that of the surrounding gas, and nozzle diameter D; hence, $D = 2r$.

The general dimensional analysis is applied to this physical phenomenon with a view to finding less complicated relations.

The diameter of the atomized liquid drops is a direct function of variables η_l, D and σ; and a reverse function of variables ρ_g, v and ρ_l. The distributed relation of $m^* = 7$ variables is

$$f_1(d,\ \rho_g,\ v,\ \rho_l) = f_2(\eta_l,\ D,\ v). \tag{24.27}$$

The dimensional matrix of variables for $d^* = 3$ fundamental quantities is:

	d	η_l	D	σ	ρ_g	v	ρ_l
L	1	−1	1	0	−3	1	−3
M	0	1	0	1	1	0	1
T	0	−1	0	−2	0	−1	0

(24.28)

The monomial relation having unknown exponents for all variables is

$$d^a\, \rho_g^b v^c\, \rho_l^e = k \eta_l^h D^i\, \sigma^j. \tag{24.29}$$

The system of linear algebraic equations expressing the dimensional homogeneity condition is:

$$\begin{aligned} a - 3b + c - 3e + h - i &= 0 \qquad \text{(L)} \\ b + e - h - j &= 0 \qquad \text{(M)} \\ -c + h + 2j &= 0 \qquad \text{(T)}. \end{aligned} \tag{24.30}$$

Bearing in mind the rule (I) applied for Diophantine systems of equations, the solution of the indeterminate system of equations is:

$$a = 3,\ b = 1,\ c = 3,\ e = 1,\ h = 1,\ i = 1,\ j = 1. \tag{24.31}$$

The relation, as expressed through a relation between physical quantities, is

$$d = k \sqrt[3]{\frac{\eta_l D \sigma}{\rho_g v^3 \rho_l}} = kR. \tag{24.32}$$

The value of the proportionality factor k is given by the results of two experimental measurements which are also checked theoretically against the data supplied by the same measurements. Using the characteristics shown in Fig. 24.1, the values are given in Table 24.1.

TABLE 24.1. **Measured values of water atomizing**

v	20	40	60	80	100	120	140	160	180	200
$10^6 d$	550	267	177	133	105	87.2	74.2	64.7	57.2	51.3

For other variables in relation (24.32) the following values are considered :

$$\begin{aligned} \eta_l &= 10.2\times10^{-6}\,\mathrm{kg\cdot s/m^2} \\ D &= 1.5\times10^{-3}\ \mathrm{m} \\ \sigma &= 7.33\times10^{-3}\ \mathrm{kg/m} \\ \rho_g &= 0.123\ \mathrm{kg.s/m^4} \\ \rho_l &= 101.7\ \mathrm{kg.s/m^4}. \end{aligned} \tag{24.33}$$

Substituting these values in relation (24.32) leads to the proportionality factor values given in Table 24.2.

As can be seen from Table 24.2, the values of the numerical factor are scattered within close enough limits : 49.3...53.1; therefore as a unique value the weighted mean value of

$$\frac{\Sigma k_n}{n} = 50.5 \tag{24.34}$$

can be admitted where n is the number of readings on the characteristic curve.

TABLE 24.2. **Computed values for the numerical factor**

$10^6 d$	550	267	177	133	105	87.2	74.2	64.7	57.2	51.3
$10^6\,kR$	10.4	5.2	3.46	2.6	2.08	1.73	1.48	1.30	1.16	1.04
k	53.1	51.1	51.0	50.8	50.6	50.2	49.8	49.6	49.5	49.3

The singular values of quantity exponents, as well as the constant value of the numerical factor, lead to the final relation :

$$d = 50.5\sqrt[3]{\frac{\eta_l D\sigma}{\rho_g v^3 \rho_l}}. \tag{24.35}$$

The values of the mean diameter of drops, as shown in Table 24.1, differ from those obtained from relation (24.35) because of the mean value adopted for the numerical factor. The differences are given in Table 24.3.

TABLE 24.3. **Comparison of measured values with those computed through the new relation**

$10^6 d$	550	267	177	133	105	87.2	74.2	64.7	57.2	51.3
$10^6 d_{(24.35)}$	523	262	175	131	105	83.0	74.6	65.4	58.2	52.4
ε, %	−4.9	−1.9	−1.1	−0.9	0	−0.5	+0.5	+1.1	+1.7	+2.1

From these results, it can be assumed that relation (24.35) is correct and consequently can be used in practice instead of a relation which implies more sophisticated computations.

If other variables arising in this physical phenomenon are taken into account, the general dimensional analysis leads to the relation :

$$d = 224\sqrt[3]{\frac{\Delta\eta D\sigma g}{\Delta\gamma v^3 \Delta\rho}} \tag{24.36}$$

where the differences between the dynamic viscosities $\Delta\eta$, the specific weights $\Delta\gamma$, and densities $\Delta\rho$ of both the liquid and the surrounding gas have been introduced. Relation (24.36) is much improved over relation (24.35) inasmuch as it takes into account all variables used to describe the phenomenon.

It is considered that relation (24.36) could be derived by solving both the systems of Navier-Stokes and of continuity equations, but this cannot be achieved as yet, owing to mathematical difficulties ; relation (24.36) should therefore be added to the theoretical study of the phenomenon [5.21].

24.3 Non-Permanent Flow in a Well

Let us consider a soil layer of a mean thickness H, above an impermeable horizontal surface. The filtering coefficient — or filtering velocity — is K while the permeability coefficient — or the movable volume of water contained within the apparent unit volume of soil, also called storage coefficient — is denoted by μ. The water in the layer has a density ρ; the flow towards the well is a non-permanent one. The co-ordinates of a point on the free surface of flow in the neighbourhood of the well are r and s (Fig. 24.2). Water is extracted from the well with a pump whose operating time is t while the constant pumping rate of flow is Q.

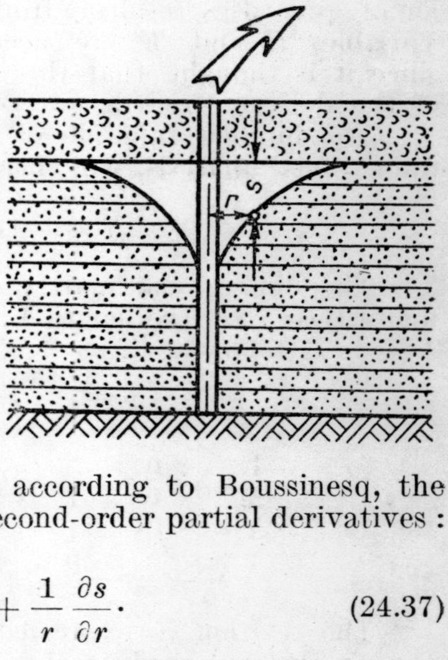

Fig. 24.2 Water flow within a well

The analytical method produces, according to Boussinesq, the following differential equation with second-order partial derivatives :

$$\frac{\mu}{HK}\frac{\partial s}{\partial t}=\frac{\partial^2 s}{\partial r^2}+\frac{1}{r}\frac{\partial s}{\partial r}. \tag{24.37}$$

According to Ferrari [5.2], the analytical solution of this problem consists in finding a particular solution $s(r, t)$ for equation (24.37) which could satisfy both the initial and the boundary conditions. Since from a mathematical point of view great difficulties are to be expected, we shall not attempt to solve the differential equation (24.37). We shall instead use dimensional analysis, which will lead us to a relation of the form $s(H, K, \mu, r, t)$; using this relation — which shall be derived in what follows — we shall construct without any difficulty whatsoever the integral of equation (24.27) and will obtain the relation

$$s = 0.183 \frac{Q}{\tau} \ln\ 2.25 \frac{t\tau}{\mu r^2} \tag{24.38}$$

where

$$\tau = HK \tag{24.39}$$

is the transmissivity.

The restricted dimensional analysis will be applied to the function of quantities resulting from the physical model described above. Variables H and K are accounted for through their product τ, since it is thought that this parameter gives a better description of the delivery from a water-containing layer.

The implicit function of $m^* = 7$ variables, used in restricted dimensional analysis, will therefore be

$$f(\mu, \rho, r, s, t, Q, \tau) = 0. \tag{24.40}$$

The dimensional matrix of variables for $d^* = 4$ fundamental quantities — i.e. the length L, the mass M, the time T and velocity V — is as follows:

	μ	ρ	r	s	t	Q	τ	
L	0	−3	1	1	0	2	2	
M	0	1	0	0	0	0	0	(24.41)
T	0	1	0	0	1	0	−1	
V	0	0	0	0	0	1	0	

The relation, as expressed by the physical quantities and the non-dimensional product of powers, is of the form

$$s = k \frac{Q}{\tau} \varphi \left(\frac{t\tau}{\mu r^2} \right) \cdot \tag{24.42}$$

In the non-dimensional product of powers, the permeability coefficient μ was introduced on the basis of its correlation with the transmissivity τ, as shown by relation (24.38). It can be seen that in relation (24.42) the water density ρ does not appear.

When integrating equation (24.37), the analytical method makes use of relation (24.42).

The general dimensional analysis may lead to a relation between the variables that describe this physical phenomenon.

Distance s varies directly with time t and with the pumping rate of flow Q, and inversely with revolving co-ordinate r (see Fig. 24.2) and with transmissivity τ. Variable μ is not taken into consideration because it has no dimensional equation.

The distributed relation of $m^* = 5$ variables is

$$f_1(r, s, \tau) = f_2(t, Q). \tag{24.43}$$

The dimensional matrix of variables for $d^* = 3$ fundamental quantities is:

	r	s	t	Q	τ
L	1	1	0	3	2
M	0	0	0	0	0
T	0	0	1	−1	−1

(24.44)

The monomial relation with unknown exponents for all variables is:

$$r^a s^b \tau^c = k t^d Q^e. \tag{24.45}$$

The system of linear equations expressing the condition of dimensional homogeneity is:

$$\begin{aligned} a + b + 2c - 3e &= 0 \quad \text{(L)} \\ -c - d + e &= 0 \quad \text{(T)}. \end{aligned} \tag{24.46}$$

The solution of the indeterminate system of equations is, according to the rule(I)applied for the resolution of Diophantine systems of equations:

$$c = 1, \quad d = 1, \quad e = 2;$$

$$\begin{cases} a = 1, & b = 3; \\ a = 2, & b = 2; \\ a = 3, & b = 1. \end{cases} \tag{24.47}$$

The matrix of the solutions is :

	a	b	d	e	c
	r	s	r	Q	τ
R_1	1	3	1	2	1
R_2	2	2	1	2	1
R_3	3	1	1	2	1

(24.48)

This result is ambiguous.

These are three relations expressed by the physical quantities, i.e. R_1, R_2, and R_3.

The value of the numerical factor of proportionality is found and checked using experimental data produced by at least two sets of measurements. In the case under consideration, no such data were available. As a consequence, for the variables to be introduced in the right-hand member of relation (24.38), usual values were adopted. Distance s is estimated with the analytical relation (24.38) with due consideration given to variable μ. The data used and the coresponding results of the calculations are presented in Table 24.4.

TABLE 24.4. **Values adopted and results of calculations by means of analytical relation**

μ	0.8	0.9	0.8	0.9
r	1.5	2.0	5.0	10.0
t	3.6×10^{5}	5.6×10^{5}	$1{,}5\times10^{6}$	2.0×10^{8}
Q	0.08	0.04	0.03	0.10
τ	3×10^{-5}	5×10^{-5}	8×10^{-5}	3×10^{-6}
$s_{(24.38)}$	239.9	79.3	33.7	3120.2

Data supplied in Table 24.4 are used to calculate relations (24.48). The value of the numerical factor of proportionality k is found by comparison. It will be seen that the value of this factor is practically constant for R_2. The result of the calculations carried out for this relation is reported in Table 24.5.

TABLE 24.5. **Computed values for the numerical factor**

$s_{(24.38)}$	239.9	79.3	33.7	3120.2
kR_2	6528	2192	919	86000
k	0.0367	0.0361	0.0366	0.0363

For the numerical factor of proportionality, a weighted mean value is accepted :

$$\frac{\Sigma k_n}{n} = 0.0364 ; \tag{24.49}$$

with this value, R_2 from the matrix of solutions (24.48) takes the final form

$$s = 0.0364 \frac{Q}{r} \sqrt{\frac{t}{\mu \tau}} . \tag{24.50}$$

The results obtained by making the same calculations with relations (24.38) and (24.50) are practically the same ; the percentage differences between the corresponding values are shown in Table 24.6.

TABLE 24.6. **Comparison between values computed with the analytical and with the new relation**

$s_{(24.38)}$	239.9	79.3	33.7	3120.2
$s_{(24.50)}$	237.6	79.8	33.4	3130.4
ε, %	+0.964	−0.631	+0.891	−0.314

For all practical uses, the author suggests relation (24.50) [5.25 ; 5.27].

24.4 Thermic Wave Damping

The thermic wave is the periodical thermal oscillation. For the sake of simplification, for all practical purposes, the periodical thermal oscillations are thought of as harmonic oscillations ; these are periodical variations produced according to either the sine or the cosine law. Periodical thermal processes are to be met in steam boiler brickwork, in the linings of industrial ovens, in the thermal insulation of piping etc. ; the periodical character of thermal processes is also important in buildings because of the daily change in the temperature of the environmental atmosphere.

As the harmonic wave penetrates the wall, it changes in amplitude, which becomes lower, and in phase, which slows down. The decrease in the amplitude of the harmonic oscillation is called damping, while the phase delay means the increase in the harmonic phase angle.

In *the analytical method* of study, complex numbers were used to permit the determination of both the amplitude and the phase angle of the harmonic under study. When the thermic wave travels from the inside towards the outside, the damping in a wall consisting of only one layer is given by the relation

$$\nu_w = 0.5\, e^{\frac{1}{\sqrt{2}} R_w s} \sqrt{1 + \frac{\alpha_e \sqrt{2}}{s} + \frac{\alpha_e^2}{s^2}}, \qquad (24.51)$$

where the symbols denote: ν_W the damping in the wall; R_w the wall thermal resistance; s the wall thermal assimilation coefficient; and α_e the coefficient of heat flow to the outside atmosphere. When the thermic wave flows from the inside air to the inside surface, damping ν_a is given by relation:

$$\nu_a = \sqrt{1 + \frac{s\sqrt{2}}{\alpha_i} + \frac{s^2}{\alpha_i^2}}, \qquad (24.52)$$

where α_i is the coefficient of heat flow to the inside atmosphere.

When the wave flows through the wall from the inside towards the outside, the damping ν is given by the product:

$$\nu = \nu_w \cdot \nu_a ; \qquad (24.53)$$

when the direction of flow is reversed, the above relation is multiplied by the ratio of heat flow coefficients $\frac{\alpha_i}{\alpha_e}$.

For situations met in practice, this relation, as established by the analytical method, involves a great amount of cumbersome calculations. This is why simplified relations were suggested — which are less general and less accurate — or else diagrammatic procedures [5.9].

The restricted dimensional analysis might yield the relation existing between the variables involved in a physical phenomenon, provided their number is not too high. From the relation above results the implicit function :

$$f(R_w, s, \alpha_e, \alpha_i, \nu) = 0. \tag{24.54}$$

The thermal resistance of a wall is estimated using the relation

$$R_w = \frac{\delta}{\lambda}, \tag{24.55}$$

where the symbols denote : ϑ wall thickness ; and λ material thermal conductivity. The wall coefficient of assimilation is

$$s = \sqrt{\frac{2\pi}{24} \cdot \frac{\lambda c \rho}{z}}, \tag{24.56}$$

where the new symbols denote : c material specific heat ; ρ material density ; z period of the oscillations. In this situation, the restricted dimensional analysis yields a relation between non-dimensional products of powers.

The general dimensional analysis might yield a simpler relation, requiring less calculation.

The study of the physical phenomenon of thermal wave damping has shown that damping ν varies in direct proportion with variables α_e, δ, c and ρ ; and in inverse proportion to variables α_i, λ and z.

The distributed relation of $m^* = 8$ variables is

$$f_1(\alpha_i, \nu, \lambda, z) = f_2(\alpha_e, \delta, c, \rho). \tag{24.57}$$

The dimensional matrix of variables for $d^* = 4$ fundamental quantities is :

	α_e	α_i	ν	δ	λ	c	ρ	z
L	0	0	0	1	1	2	−3	0
M	1	1	0	0	1	0	1	0
T	−3	−3	0	0	−3	−2	0	1
Θ	−1	−1	0	0	−1	−1	0	0

(24.58)

The monomial relation with unknown exponents to all variables is:

$$\alpha_i^a \nu \lambda^b z^d = k \alpha_e^e \delta^g c^h \rho^i. \qquad (24.59)$$

The system of linear equations expressing the condition of dimensional homogeneity is:

$$\begin{aligned} b - g - 2h + 3i &= 0 \qquad (L)\\ a + b - e - i &= 0 \qquad (M)\\ -3a - 3b + d + 3e + 2h &= 0 \qquad (T)\\ -a - b + e + h &= 0 \qquad (\Theta). \end{aligned} \qquad (24.60)$$

The solution for the indeterminate system of equations, according to the rule (I) for the resolution of Diophantine systems of equations, is:

$$a = 1,\ b = 1,\ e = 1,\ i = 1,\ h = 1,\ g = 2,\ d = 1. \qquad (24.61)$$

The relation as expressed by the physical quantities is

$$\nu = \frac{\alpha_e \delta^2 c \rho}{\alpha_i \lambda z} = kR. \qquad (24.62)$$

The value of the numerical factor of proportionality k is derived and checked using at least 2 sets of experimental measurements. But in this case no data were available. As a consequence, for the variables included in relation (24.62), the usual values were adopted, i.e.

$$\begin{aligned} \alpha_e &= 20\ \text{kcal/m}^2 \cdot \text{h} \cdot {}^\circ\text{C}\\ \alpha_i &= 7\ \text{kcal/m}^2 \cdot \text{h} \cdot {}^\circ\text{C}\\ \lambda &= 0.65\ \text{kcal/m} \cdot \text{h} \cdot {}^\circ\text{C}\\ c &= 0.20\ \text{kcal/kg} \cdot {}^\circ\text{C}\\ \rho &= 1.700\ \text{kg/m}^3. \end{aligned} \qquad (24.63)$$

With these values, and ascribing variable values to δ and z, damping ν is estimated by using the relation derived by the analytical method. The results of the calculations are recorded in Table 24.7.

Data tabulated in Table 24.7 are used to make the computations implied by relation (24.62). A comparison of data obtained will yield the values of the numerical factor of proportionality k. The results of the calculations are reported in Table 24.8.

TABLE 24.7. **Values adopted and those computed with the analytical relation**

δ	0.320				0.425					0.550					
R_w	0.493				0.655					0.845					
z	1.0	1.3	1.5	1.7	1.8	2.1	2.3	2.6	3.0	3.0	3.3	3.5	4.0	4.1	4.5
s	7.60	6.66	6.20	5.83	5.66	5.25	5.02	4.71	4.38	4.38	4.18	4.06	3.80	3.75	3.58
ν	48.7	35.8	31.3	28.2	50.6	43.4	39.2	35.2	31.2	56.0	50.6	47.2	42.3	41.4	38.3

TABLE 24.8. **Values computed for the numerical factor**

$\nu_{(24.53)}$	48.7	35.8	31.3	28.2	50.6	43.4	39.2	35.2	31.2	56.0	50.6	47.2	42.3	41.4	38.3
kR	153.4	118.4	102.1	89.7	150.4	129.0	118.3	103.7	90.5	150.4	137.1	129.0	112.6	110.2	100.4
k	0.318	0.302	0.310	0.314	0.336	0.338	0.331	0.340	0.345	0.372	0.369	0.366	0.376	0.376	0.380

As can be seen from the table above, for a wall whose thickness is δ, the value of the numerical factor varies between very narrow boundaries; for walls of other thicknesses, the oscillations take other values whose changes are independent of the values adopted for the oscillation period; the random values ascribed to z affect only the values found for the numerical factor. Making use of Table 24.8, weighted mean values are adopted for the numerical factor k_δ. The calculations yield : $k_{0.320} = 0.310$; $k_{0.425} = 0.338$ and $k_{0.550} = 0.379$. With these values, relation (24.62) takes the final form

$$\nu = k_\delta \frac{\alpha_e \delta^2 c \rho}{\alpha_i \lambda z}. \tag{24.64}$$

It is assumed that the value of the numerical factor is

$$k_\delta = f(\delta). \tag{24.65}$$

The mean values adopted can be represented by means of a system of rectangular axes where the wall thickness δ is on the abscissa, while the numerical factor of proportionality k, on the ordinate axis. For two points, the following equations could be written :

$$\begin{aligned} a_1 &= \mathrm{tg}\varphi_1 = \frac{k_2 - k_1}{\delta_2 - \delta_1} = 0.266 \\ a_2 &= \mathrm{tg}\varphi_2 = \frac{k_3 - k_1}{\delta_3 - \delta_1} = 0.274\,; \end{aligned} \tag{24.66}$$

hence the representative curve will actually be a straight line having a slope $a = 0.27$. At the origin, the value of the ordinate is $b = 0.23$ as measured on the diagram. Hence, the representative straight line is

$$k = 0.27\ \delta + 0.23. \tag{24.67}$$

Consequently, in its final form, the relation describing the damping ν will be

$$\nu = \left(0.27 \frac{\delta}{\delta_0} + 0.23\right) \frac{\alpha_e \delta^2 c \rho}{\alpha_i \lambda z}, \tag{24.68}$$

where $\delta_0 = 1$.

Calculations carried out with relations (24.53) and (24.68) will yield practically the same results ; the differences are reported, in percentage values, in Table 24.9.

TABLE 24.9. **Comparison between values computed with the analytical and with the new relations**

$\nu_{(24.53)}$	48.7	35.8	31.3	28.2	50.6	43.4	39.2	35.2
$\nu_{(24.68)}$	47.6	36.8	31.6	27.8	50.8	43.6	39.8	35.0
ε, %	−2.2	+3.0	+0.9	−1.4	+0.4	+0.5	+1.5	−0.6
		31.2	56.0	50.6	47.2	42.3	41.4	38.3
		30.8	56.1	51.1	48.0	42.0	41.0	37.3
		−1.3	+0.2	+1.0	+1.7	−0.7	−1.0	−2.6

For other materials, the quantities from relation (24.68) can take other values than those listed above. Heat transfer coefficients differ with the nature of the phenomenon. The oscillation period is variable too. In this situation, the calculations can be carried out as above and a diagram can be drawn up for the variables

$$\nu = f(R_w, s, \delta) \tag{24.69}$$

which will be used for the determination of damping values for all practical purposes [5.20].

24.5 Stationary Convection in Forced, Turbulent Air Flow

Assuming that a heat conducting solid body of a given temperature is immersed into a liquid flow having a different temperature, we shall then be in the presence of a heat transfer process called heat convection. This physical phenomenon will be studied by assuming that the solid body is a plate of length l while the width has a unit value. The fluid is air of a density ρ, a dynamic viscosity η, and a specific heat c; the velocity of flow is v; the differential temperature is $\Delta\theta$. According to Newton, the heat amount exchanged in the unit time can be estimated with relation $Q = \alpha\, l\, \Delta\theta$, where the new symbol represents the coefficient of heat transfer α.

Even before 1920, Stanton[24.2] applied *the analytical method* to the study of this physical phenomenon. He considered the fluid jet as a whole, therefore neglecting the boundary layer. The result obtained is shown in the relation:

$$\alpha = 3.01\,(1 + 0.0075\,\Delta\theta)v^{0.75}. \tag{24.70}$$

[24.2] Stanton, T. E., Friction, *Rep. and Mem.*, No. 271, April 1917.

In 1920, Kármán Latzko[24.3] studied this physical phenomenon on the basis of the similarity between the mechanism of friction and that of heat transfer phenomena. The results obtained [5.4] are shown in the relation :

$$\alpha = 128.16\, l^{-0.2} \nu^{0.2} c v^{0.8}. \tag{24.71}$$

The experimental method was applied by several researchers. The experimental measurements were carried out for the air at a temperature of 40°C by Doetch[24.4]. The conditions were as follows :

$$\begin{aligned} l &= 0.5 \text{ m} \\ \nu &= 17.5 \times 10^{-6} \text{ m}^2/\text{s} \\ c &= 0.26 \text{ kcal/m}^3.°\text{C}. \end{aligned} \tag{24.72}$$

Throughout the measurements, the air jet did not exceed a velocity of 70 to 100 m/s. The results are diagrammatically illustrated in a system of rectangular axes, the air jet velocity being represented on the abscissa, with the coefficient of heat transfer, on the ordinate ; the diagram drawn up for the experimental data is shown in Fig. 24.3. For comparison purposes, on the same diagram the data resulting from the relations developed by Stanton and by Kármán-Latzko are also plotted.

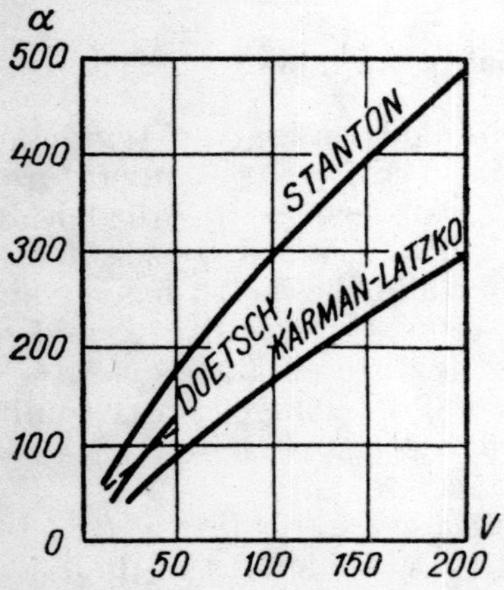

Fig. 24.3 Curves expressing the analytical relations, the experimental measurements and the new relation

[24.3] Kármán, Th. V., Über laminäre und turbulente Reibung, *Abh. Aerod. Inst. Aachen* no. 1, 1920.

Latzko, H., Der Wärmeübergang an einer turbulenten Flussigkeitsoder Gasstrom, *Abh. Aerod. Inst. Aachen*, No. 1, 1920.

[24.4] Doetsch, H., Die Wärmeübertragung von Kühlrippen an strömende Luft, *Aerod. Inst. Techn. Hochschule Aachen*, no. 14, 1934.

As can be seen, the measured values of the heat transfer coefficient lie in the middle, between the values yielded by Stanton's relation (which are higher) and those yielded by Kármán-Latzko's (which are lower). Therefore, the analytical relations are not checked by the experimental data [5.4].

The restricted dimensional analysis can be applied to the function of $m^* = 5$ variables included in the relation suggested by Kármán-Latzko, namely

$$f(l, \nu, c, v, \alpha) = 0. \tag{24.73}$$

This method will lead to a relation between non-dimensional products of powers ; a form of this relation can be identical with relation (24.71).

The general dimensional analysis might lead to a relation that will agree well with the experimental data reported by Doetsch.

An investigation into the mechanics of the physical phenomenon will show that the coefficient of heat transfer α varies in direct proportion with the kinematic viscosity ν, the specific heat c and the velocity v; and in inverse proportion to the length l of the plate. The distributed relation of $m^* = 5$ variables is

$$f_1(l, \alpha) = f_2(\nu, c, v). \tag{24.74}$$

The dimensional matrix for $d^* = 4$ fundamental quantities is :

	l	ν	c	v	α
L	1	2	−1	1	0
M	0	0	1	0	1
T	0	−1	−2	−1	−3
Θ	0	0	−1	0	−1

(24.75)

The monomial relation with unknown exponents for all variables is

$$l^a \alpha^b = k \nu^d c^e v^g. \tag{24.76}$$

The system of linear equations expressing the condition of dimensional homogeneity is :

$$\begin{aligned} a - 2d + e - g &= 0 \quad (\mathrm{L}) \\ b - e &= 0 \quad (\mathrm{M}) \\ -3b + d + 2e + g &= 0 \quad (\mathrm{T}) \\ -b + e &= 0 \quad (\Theta). \end{aligned} \tag{24.77}$$

The solution of the indeterminate system of equations, according to the rule (I) applied to the resolution of Diophantine systems of equations, is:

$$b = 2, \quad d = 1, \quad a = 1, \quad e = 1, \quad g = 1. \tag{24.78}$$

The relation, as expressed by the physical quantities is

$$\alpha = k\sqrt{\frac{\nu c^2 v}{l}}. \tag{24.79}$$

The value of the proportionality factor k will then be found and checked according to the theory, using the results from two sets of experimental measurements. For this purpose the experimental data reported by Doetsch will be used. Data presented in Table 24.10 are taken from Fig. 24.3.

Making use of the experimental data given in (24.72) and in Table 24.10, relation (24.79) will yield the values of the numerical factor of proportionality k. However, these values are found to vary within wide boundaries and consequently it follows that relation (24.79) is not correct. Hence, for the quantity exponents a solution involving integral, small, positive values must be adopted.

TABLE 24.10. **Values measured for air convection**

v	10	13	16	19	24	29	34	43
α	36	44	52	60	72	83	95	110

The solution of the indeterminate system of equations (24.77) is found by applying the rule (II) for the resolution of Diophantine systems of equations. The solution matrix is:

$$\begin{array}{c|ccccc|} & a & d & e & g & h \\ & l & \nu & c & v & \alpha \\ \hline R_1 & 1 & 1 & 3 & 2 & 3 \\ R_2 & 2 & 2 & 3 & 1 & 3 \\ R_3 & 1 & 1 & 4 & 2 & 4 \\ R_4 & 2 & 2 & 4 & 2 & 4 \\ R_5 & 3 & 3 & 4 & 1 & 4 \end{array} \tag{24.80}$$

The relation, as expressed by the physical quantities, is written for each line of values from the solution matrix (24.80), i.e. $R_1, R_2, R_3, \ldots$

For the relations above, the values of the numerical factor of proportionality k will be computed with the same experimental data. The relation where the value of the numerical factor is constant is the correct one. For R_3 these values will vary within narrow boundaries; we therefore present, in Table 26.11 below, the results of the calculations carried out for this relation only.

TABLE 24.11. **Values computed for the numerical factor**

α	36	44	52	60	72	83	95	110
kR_3	0.114	0.138	0.162	0.183	0.220	0.254	0.284	0.339
k	317	317	321	327	327	327	333	325

Making use of the values listed in Table 24.11, we shall adopt the weighted mean value

$$\frac{\Sigma k_n}{n} = 325 \tag{24.81}$$

and relation R_3 will take the final form

$$\alpha = 325 \sqrt{\frac{\nu c^4 v^3}{l}}. \tag{24.82}$$

Data listed in Table 24.12 below are intended to afford a comparison between the experimental results reported by Doetsch and the data calculated with relation (24.82). The table lists the differences, in percentages, between the two sets of values. For a better illustration of the correlation between the two sets of values, data listed in Table 24.12 are diagrammatically represented in Fig. 24.3. As can be seen, data reported by Doetsch are in good agreement with those calculated with relation (24.82), as derived through general dimensional analysis [5.14].

TABLE 24.12. **Comparison between values measured and those computed with the new relation**

α	36	44	52	60	72	83	95	110
$\alpha_{(24.82)}$	37.1	45.2	52.9	59.7	71.8	82.9	92.8	109.6
ε, %	+2.9	+2.6	+1.7	−0.5	−0.3	−0.1	−2.3	−0.4

24.6 Heat Convection through a Porous Medium

Heat transfer by natural convection through a porous medium is a geothermal problem: it appears in the motion of underground water and in the mechanism of volcanic phenomena; it is also of interest for insulating materials and for the highly porous materials, used in nuclear and in cryogenic technologies. Nowadays, this physical phenomenon is the focus of great concern and is the subject of a great deal of experimental and theoretic investigation.

The analytical method of study was applied by Horton and Rogers in 1945[24.5] and, separately, by Lapwood, in 1948[24.6]. These authors make use of the equations and methods applied in describing the heat transfer through a horizontal layer of viscous fluid; the viscosity forces from the Navier-Stokes equations are replaced by the filtering forces, while the local velocities are replaced by the filtering velocities. It is, however, worth mentioning that other authors have lately shown an interest in the matter [5.5].

In writing the general equations for the mass, the amount of motion and the energy conservation, due consideration was given to the following:

—the solid material is non-deformable, homogeneous and isotropic;

—the interstitial fluid has but one phase and saturates the medium;

— the interstitial fluid average temperature is equal to that of the solid material;

Now, also bearing in mind that, the heat transfer conditions are stationary, the filtering velocity is low enough (the hypothesis of the linear theory), and the porous medium is a two-dimensional one, then the equations mentioned above will be written in the form:

$$\frac{\partial(\rho u)}{\partial x}+\frac{\partial(\rho v)}{\partial y}=0$$

$$u=\frac{K}{\eta}\left(g\rho-\frac{\partial p}{\partial x}\right)$$

$$v=-\frac{K}{\eta}\frac{\partial p}{\partial y} \tag{24.83}$$

$$u\frac{\partial}{\partial x}[(\rho c)_f\,\theta]+v\frac{\partial}{\partial y}[(\rho c)_f\,\theta]=\frac{\partial}{\partial x}\left(\lambda^*\frac{\partial\theta}{\partial x}\right)+\frac{\partial}{\partial y}\left(\lambda^*\frac{\partial\theta}{\partial y}\right).$$

24.5 Horton, C. W., Rogers, F. T., Convection currents in a porous medium, *J. Appl. Phy.*, XVI, No. 6, 1945, p. 367–370.

24.6 Lapwood, E. R., Convection of a fluid in a porous medium, *Proc. Cambr. Phil. Soc.*, XXXXIV, 1948, p. 508–521.

To these, one should add the state equation of the interstitial fluid

$$\rho = \rho_0[1 - \alpha(\theta - \theta_0)] \tag{24.84}$$

as well as other functions expressing the change with temperature in the values of quantities C, η and λ^*. The symbols above denote : ρ — density of the interstitial fluid ; u and v — components of the filtering velocity following the x and y axes respectively ; K — permeability of the solid material ; g — normal acceleration in free fall ; p — pressure ; η — fluid dynamic viscosity ; c — specific heat at a constant pressure ; θ — temperature ; λ^* — apparent thermal conductivity of the porous medium saturated with motionless interstitial liquid ; α — coefficient of fluid expansion. Subscript f refers to the fluid.

The accurate solution of the system of differential equations (24.83) has not, as yet, been found. The system was numerically integrated on a computer ; the solutions were convergent up to Reynolds numbers of about 10^3. The approximate integration led to the solution :

$$\overset{*}{Nu} = k \sqrt{\frac{\overset{*}{Ra}}{A}} \tag{24.85}$$

whose main components are the Nusselt's invariant for the porous medium ; and the Rayleigh's invariant for the porous medium, as given by the relation :

$$\overset{*}{Ra} = g \frac{\alpha(\rho c)_f}{\nu} \cdot \frac{K}{\lambda^*} \Delta\theta\, d \tag{24.86}$$

and the ratio

$$A = \frac{L}{H} \cdot \tag{24.87}$$

The new symbols in relations (24.86) and (24.87) represent the differential temperature between the hot wall and the cold one $\Delta\theta$, the mean diameter (of either the grain or the fibre) d, the kinematic viscosity of the fluid ν, the height of the vertical cell containing the porous medium L, and the distance between the hot and the cold walls of cell H.

Data obtained by numerical integration were used to determine the value of k, and the following half-empirical relation was obtained:

$$N\overset{*}{u} = 0.55 \sqrt{\frac{R\overset{*}{a}}{A}} \cdot \mathrm{th}(0.575\ A). \tag{24.88}$$

This relation is valid in principle only for separate limited layers and makes sense only if the second number has a unit or above-unit value. In deriving relation (24.88), the following values were used [5.5]:

$$10^{-1} < \frac{R\overset{*}{a}}{A} < 10^{3}$$
$$0.125 < A < 100 \tag{24.89}$$

The experimental method was applied for a vertical parallelepiped cell of solid material and for the conditions shown in Table 24.13.

In Table 24.13 the new symbols represent: ρ_a — the solid phase apparent density; ε — porosity. The measurement results are tabulated too.

Phenyl chloride was used as interstitial fluid, the technical data of this fluid being:

$$\begin{aligned} \alpha &= 0.985 \cdot 10^{-3}\ {}^\circ\mathrm{C}^{-1}\ (\text{between } 0 \text{ and } +30^\circ\mathrm{C}); \\ c &= 0.308\ \mathrm{kcal/kg} \cdot {}^\circ\mathrm{C}\ (\text{at } +20^\circ\mathrm{C}); \\ \rho &= 1219\ (1 - 0.812 \cdot 10^{-3}\ \theta)\ \mathrm{kg/m^3}; \\ \eta &= 0.803 \cdot 10^{-3}\ (273 + \theta)/293\ \mathrm{P}; \\ \lambda &= 0.115 + 1.978 \cdot 10^{-4}\theta\ \mathrm{kcal/m \cdot h \cdot {}^\circ C}. \end{aligned} \tag{24.90}$$

The restricted dimensional analysis can be applied to the function of variables that results from the equations above; the result obtained is an ill-defined one, including a function of invariants that cannot be determined by this method.

The general dimensional analysis will show that this physical phenomenon can be described by means of a simpler, and at the same time more accurate, relation.

TABLE 24.13. **Conditions of solid materials and results obtained**

Test No.	Porous material	θ_1	θ_2	A	$\overset{*}{Ra}$	$\overset{*}{Nu}$
1	Glass fibre $\rho_a = 160$ kg/m³ $\varepsilon = 0.936$ $d = 6$ μm $K// = 180\ 10^{-12}$ m²	16.0	54.0	4.50	44.2	1.5
2 3 4 5 6 7	Glass fibre $\rho_a = 70$ kg/m³ $\varepsilon = 0.972$ $d = 6$ μm $K// = 760\ 10^{-12}$ m²	17.0 12.6 15.0 15.0 13.0 10.0	22.0 22.4 30.0 35.0 40.0 45.0	2.25	47.9 91.6 147.6 200.6 252.4 280.0	2.2 3.0 3.8 4.5 5.2 6.0
8 9 10 11 12	Glass fibre $\rho_a = 40$ kg/m³ $\varepsilon = 0.984$ $d = 12$ μm $K// = 5{,}200\ 10^{-12}$ m²	16.0 18.0 15.0 15.0	21.0 28.0 35.0 40.0	2.25	324.5 649.0 1,298.0 1,622.5	5.7 8.3 11.8 13.3
12 13 14	Glass fibre $\rho_a = 40$ kg/m³ $\varepsilon = 0.984$ $d = 12$ μm $K// = 5{,}200\ 10^{-12}$ m²	18.0 18.0 15.0	23.0 28.0 35.0	4.50	162.0 324.0 648.0	3.3 4.7 6.8
15	Glass powder $\varepsilon = 0.400$ $630 < d < 800$ m $K = 588\ 10^{-12}$ m²	25.0	60.0	4.50	37.9	1.4

The Nusselt number for a porous medium is defined by means of the relation

$$\overset{*}{Nu} = \frac{\lambda_a}{\lambda^*}; \tag{24.91}$$

then it is demonstrated that this number is a function of only two parameters [5.5], i.e.

$$\overset{*}{Nu} = f(\overset{*}{Ra},\ A). \tag{24.92}$$

Looking at Table 24.13, we can see that in experiments 2–7 the following data stay constant: ratio A, diameter d and permeability K. Hence, the experiments show the effects of temperature difference $\Delta\theta$; thus

$$\overset{*}{Nu} = kf(\Delta\theta) = k\Delta\theta^{x}. \tag{24.93}$$

The results reported for experiments 8–11 show that value A is maintained as above, while the mean diameter and the permeability values step higher; for the same values of $\Delta\theta$ will be easily found that:

$$\overset{*}{Nu} = kf(d, K) = kd^{y}K^{z}. \tag{24.94}$$

Experiments 8–11 and 12–14 are of the same type but differ by the value of ratio A; the effect of this variable being thus proved, it follows that:

$$\overset{*}{Nu} = kf(A) = kA^{-w}. \tag{24.95}$$

From relations (24.86), (24.94) and (24.95) it follows that relation (24.92) takes the form:

$$\overset{*}{Nu} = k\overset{*}{Ra}{}^{t}A^{-w}. \tag{24.96}$$

Bearing in mind relations (24.86), (24.87) and (24.91), it is obvious that relation (24.91) can also be written in the form

$$\frac{\lambda_a}{\lambda^*} = k\,\frac{f_1(g, K, \alpha, d, \Delta\theta, c, \rho, H)}{f_2(\nu, \lambda^*, L)}, \tag{24.97}$$

the quantity λ^* being contained in both members.

The correct relation with $m^* = 12$ variables distributed between the two members is

$$f_1(\lambda_a, \nu, L) = kf_2(g, K, \alpha, d, \Delta\theta, c, \rho, H, \lambda^*). \tag{24.98}$$

The dimensional matrix of variables for $d^* = 5$ fundamental quantities is:

	g	K	α	d	$\Delta\theta$	c	ρ	H	λ_a	λ^*	ν	L	
L	1	2	0	1	0	0	−3	1	−1	−1	2	1	
M	0	0	0	0	0	−1	1	0	0	0	0	0	
T	−2	0	0	0	0	0	0	0	−1	−1	−1	0	(24.99)
Θ	0	0	−1	0	1	−1	0	0	−1	−1	0	0	
Q	0	0	0	0	0	1	0	0	1	1	0	0	

The monomial relation with unknown exponents for all variables is

$$\lambda_a^b \nu^h L^i = k g^j K^l \alpha^m d^n \Delta\theta^o c^p \rho^q H^r \lambda^{*s}. \qquad (24.100)$$

The system of linear equations expressing the condition of dimensional homogeneity is:

$$\begin{aligned} -b + 2h + i - j - 2l - n + 3q - r + s &= 0 && (L) \\ p - q &= 0 && (M) \\ -b - h + 2j + s &= 0 && (T) \quad (24.101) \\ -b + m - o + p + s &= 0 && (\Theta) \\ b - p - s &= 0 && (Q). \end{aligned}$$

The solution of the indeterminate system of $e^* = 5$ equations with $n^* = 12$ unknowns will be, according to the rule (I) applied to the resolution of Diophantine systems of equations:

$$\begin{aligned} &h = i = j = l = m = n = o = p = q = r = s = 1 \\ &b = 2. \end{aligned} \qquad (24.102)$$

The relation, as expressed by the physical quantities, is

$$\lambda_a^2 = k \frac{g K \alpha d \Delta\theta (c\rho)_f H \lambda^*}{\nu L}. \qquad (24.103)$$

Bearing in mind relations (24.86) and (24.87), and dividing relation (24.103) by λ^* the relation between the invariants results in the form

$$\overset{*}{Nu} = k \sqrt{\frac{\overset{*}{Ra}}{A}}. \qquad (24.104)$$

Now, using the data given in Table 24.13, the values of the non-dimensional numerical factor k are calculated and the results are shown in Table 24.14.

TABLE 24.14. **Values of numerical factor**

Test No.	A	$\overset{*}{Ra}$	$\overset{*}{Nu}$	k
1	4.50	44.2	1.5	0.477
2	2.25	47.9	2.2	0.478
3		91.6	3.0	0.472
4		147.6	3.8	0.472
5		200.6	4.5	0.475
6		252.4	5.2	0.490
7		280.0	6.0	0.536
8	2.25	324.5	5.7	0.475
9		649.0	8.3	0.486
10		1,298 0	11.8	0.492
11		1,622.5	13.3	0.497
12	4.50	162.0	3.3	0.550
13		324.0	4.7	0.552
14		648.0	6.8	0.566
15	4.50	37.9	1.4	0.489

The results obtained show that the values of factor k vary within narrow boundaries for a given value of ratio A; the following average values can be accepted:

$$k = 0.485 \text{ for } A = 2.25$$

$$k = 0.555 \text{ for } A = 4.50. \qquad (24.105)$$

The relation between k and A is written in the form:

$$k = k' A^x. \qquad (24.106)$$

Using the k and A values from Table 24.14, we obtain:

$$k' = 0.415$$

$$x = 0.195. \qquad (24.107)$$

Relation (24.104) takes the form:

$$N\overset{*}{u} = 0.415\, A^{0.195} \sqrt{\frac{R\overset{*}{a}}{A}} \tag{24.108}$$

Considering that:

$$k = k_1 + k_2\, A \tag{24.109}$$

the following simpler relation is obtained:

$$N\overset{*}{u} = (0.412 + 0.032\, A) \sqrt{\frac{R\overset{*}{a}}{A}}\, . \tag{24.110}$$

The relation above, derived on a dimensional basis, agrees well with experimental data; indeed, when compared to the half-empirical relation (24.88) obtained by applying the analytical study, the relation established by the general dimensional analysis of either forms, (24.108) or (24.110), yields values closer to data obtained by measurements.

The physical relations above were derived by the author by general dimensional analysis.

★

In the work [5.1] it is admitted that relation

$$f(\alpha, d_e, \nu) = f_2\, (v, \rho, c_p, \lambda) \tag{24.111}$$

describes the coefficient of thermal convection α; other symbols denote: d_e — equivalent diameter of the conduit; ν — kinematic viscosity of the fluid; v — velocity of flow; ρ — fluid density; c_p — fluid specific heat; and ν — fluid thermal conductivity.

According to the general method of dimensional analysis, unknown exponents are ascribed to all variables and the equations expressing the condition of dimensional homogeneity of relation (24.111) are written down; the quantity exponents can be found by applying the theorem of positive, small, integral values; then the relation

$$\alpha^5 = k\, \frac{v^4 \rho^2 c_p^2 \lambda^3}{\nu^2 d_e} \tag{24.112}$$

where k is a non-dimensional numerical factor, is considered. The elliptic shape of the conduit cross-section is accounted for by introducing in the above relation the ratio of semi-axes a and b.

From experimental data, the final relation is obtained :

$$\alpha = 0.023\, v^{0.8} \left(\frac{\rho c_p}{\nu}\right)^{0.4} \lambda^{0.6} d_e^{-0.2} \left(\frac{b}{a}\right)^{2.1} - 0.48 \log Re \quad (24.113)$$

where the Reynolds number is also included.

★

In 1979, the method of the general dimensional analysis was applied[24.7] to describe the erosion of nests of tubes of steam boiler economizers.

Erosion describes that physical phenomenon involving the tear-off of fragments from a material as a consequence of the impact of solid particles; the effects of the phenomenon are reduced sizes and changed shape of the body subjected to erosion. This phenomenon takes place in steam boilers fired with solid fuels, where the nests of tubes are swept by the ash-carrying burning gases; the economizers are the components most exposed to this phenomenon.

An investigation into the erosion phenomenon will show that on tubes, the maximum tube-wall thinning, h_{max}(cm), depends on the following physical quantities: M (gf/sq.cm.s) — the specific, gravimetric ash flow-rate; γ_c(gf/cu.cm) — ash specific weight; v (cm/s) — velocity at the moment when the ash impacts tube-wall; d_p(cm) — ash particle mean diameter; E (gf/sq.cm) — modulus of elasticity of tube material; H (gf/sq.cm) — hardness of tube material; d_t(cm) — tube OD; ν (sq.cm/s) — kinematic viscosity of burning gases; γ_g(gf/cu.cm) — specific weight of burning gases; τ (s) — time duration of the process. Thus, the following relation can be written :

$$h_{max} = \frac{f_1(M, \gamma_c, v, d_p, E, \tau)}{f_2(H, d_t, \nu, \gamma_g)} \quad (24.114)$$

where due consideration was paid to the correlation between the variables.

Tube erosion is therefore a complex physical phenomenon the description of which requires a comparatively high number of variables. The analytic method has failed, as yet, to yield a relation including the quantities from relation (24.114).

24.7 D. V. Pillat, *A relation for estimating the ash-erosion level in economizer tubes.* Energetica, XXVIII, No. 12, 1980, pp. 547—550.

The erosion phenomenon was investigated in operational conditions specific to economizers. The first set of tests was conducted on a nest of tubes placed in an open-loop wind tunnel; the corrosive agent was the ashes from the thermoelectric power stations while the carrying fluid was air at the environmental temperature. The second set of tests had the object of following the erosion effects on tube samples placed in the flues of the economizer areas of two steam boilers; the parameters intervening in this physical phenomenon were measured.

Lacking any analytical relation, we give below the form of relation (24.114) which is derived through other methods of general dimensional analysis.

Bearing in mind the results reported from theoretical and experimental investigations conducted on the erosion phenomenon, it follows that

$$h_{\max} = Kv^2\tau, \tag{24.115}$$

where K accounts for the remaining independent variables. As a consequence, according to the general dimensional analysis, relation (24.114) will be written

$$h^a_{\max} H^b d^e_t \nu^f \gamma^h_g = kM^i \gamma^j_c v^2 d^l_p E^m \tau, \tag{24.116}$$

where k is a non-dimensional numerical factor.

The system of $e^* = 3$ algebraic non-homogeneous linear equations with $n^* = 9$ unknowns is

$$\begin{aligned} a-2b+e+2f-3h &= -2i - 3j + 2 + l - 2m \quad &(\mathrm{L})\\ b+h &= i + j - m &(\mathrm{F})\\ -f &= -i + 1 &(\mathrm{T}), \end{aligned} \tag{24.117}$$

expressing the condition of dimensional homogeneity for relation (24.116), is an indeterminate one; the new symbols represent fundamental quantities, length L, force F, time T.

The theorem of positive, integral, minimum values leads to two solutions; the values of unknowns are represented in the following matrix of solutions:

	a	b	e	f	h	i	j	l	m
R_1	1	2	1	2	1	1	1	4	1
R_2	1	1	1	2	2	1	1	3	1

(24.118)

Two relations have resulted therefore, i.e. R_1 and R_2, both dimensionally homogeneous.

The value of the non-dimensional numerical factor, as estimated on the basis of experimental data, is maintained practically constant only in the relation expressing the second solution ; the values obtained were contained within the interval :

$$k = 4 \times 10^{-17} \ldots 5 \times 10^{-17}. \tag{24.119}$$

The arithmetic mean of these values is introduced in the relation :

$$h_{\max} = 4.42 \cdot 10^{-17} \frac{M \gamma_c v^2 d_p^3 E \tau}{H d_t \nu^2 \gamma_g^2} \tag{24.120}$$

which is the expression of the maximum thinning (erosion) of an economizer tube-wall as a result of the action of burning gases.

25 Physical Relations Checked

25.1 Relation Checked

Poisseuille derived through the analytical method a relation for the following physical phenomenon : a volume V of a liquid having the dynamic viscosity η flows in a time t through a pipe of radius r and length l, under a pressure drop Δp, according to relation [5.2] :

$$V = k \frac{r^4 \Delta p t}{l \eta}. \tag{25.1}$$

The distributed relation of $m^* = 6$ variables :

$$f_1 (l, V, \eta) = f_2 (r, \Delta p, t). \tag{25.2}$$

The monomial relation with unknown quantity exponents for all variables is :

$$l^a V^b \eta^c = k r^d \Delta p^e t^g. \tag{25.3}$$

The relation expressed through the physical quantities, as yielded by the general dimensional analysis, is (25.1) and can be readily derived.

25.2 Relations that do not Check

Let us consider the physical phenomenon of fluid boiling. The heat transfer always begins from the heated surface. Boiling — hence vaporization — could be started by the roughness of the vessel's inner surface, by the gas particles developed within the liquid, or else by the ionized liquid particles produced within the liquid, when sufficiently overheated with respect to the saturation temperature. For a smooth surface, boiling can be started at a liquid temperature equal to that of the rough surface, provided a few grains of sand are poured into the vessel. The heat stream-lines are concentrated at the tips of the rugged excrescences and, as a consequence, vapour bubbles can be generated. These grow with the heat input and, on reaching a certain size, become detached from the heating surface and start rising. In the case of water, for instance, on traversing the liquid mass, the bubbles grow in size since the water is hotter than its saturation temperature while the pressure produced on it by the water column continously drops.

Let us assume that the liquid has a thermal conductivity λ, a thermal diffusivity a and a specific weight γ_l, while the vapour specific weight is γ_v. The vapour bubble has a diameter d and a lift velocity through the water layer — v. The normal acceleration in free fall is g. It is necessary to find the relation for the coefficient of heat transfer α. The specific weight differential is noted by $\Delta\gamma$.

Water boiling in natural convection conditions was studied by E. Ruckenstein [5.8], who investigated the heat transfer from the liquid to the moving bubbles on a physical model, according to which the heat exchange takes place between small bubbles and minuscule amounts of liquid that come into contact with the bubbles for a short time; this model leads to the following relation for the heat transfer :

$$\alpha = \sqrt{\frac{4\lambda^2 v}{\pi a d}}. \tag{25.4}$$

For Reynolds numbers from 400 to 1800, the velocity v is given by the relation

$$v = \sqrt{\frac{4dg\,\Delta\gamma}{3\zeta\,\gamma_l}}, \tag{25.5}$$

where ζ is a friction factor whose value is about 0.6. As can be readily seen, relation (25.4) is a homogeneous one. It yields values of the same order of magnitude as those reported for the experimental measurements carried out by Fritz and Ende [5.8].

From relations (25.4) and (25.5), we obtain the function of $m^* = 7$ variables involved in this physical phenomenon :

$$f(\lambda, a, \gamma_l, d, g, \alpha, \Delta\gamma) = 0. \tag{25.6}$$

The distributed relation with unknown exponents for all variables is

$$a^b \gamma_l^c d^e \alpha^h = k \lambda^i g^j \Delta\gamma^m. \tag{25.7}$$

As can be seen, the value derived for a quantity exponent through general dimensional analysis is negative, namely $e = -2$. The result is opposed to the variable distribution, as originally adopted on the ground of relation (25.4); furthermore, the value of the quantity exponents differ.

Ruckenstein [5.8] also suggests a relation for the change in the vapour bubble diameter with distance X from the place where the latter was generated. Other variables are : the bubble original diameter d_0, the latent heat of vaporization r and the temperature difference $\Delta\theta$. The relation suggested is of the form

$$d^{7/4} - d_0^{7/4} = \frac{7\lambda X \Delta\theta}{\gamma_v r} \sqrt[4]{\frac{3\zeta\gamma_l}{4\pi^2 a^2 g \Delta\gamma}}. \tag{25.8}$$

The general dimensional analysis calculus conducted on the function of variables from relation (25.8) leads in this case again to a negative value for one quantity exponent and to different values for the quantity exponents [5.23].

It is therefore the opinion of the author that for such situations the analytical method cannot yield complete results. This conclusion is suggested by the general dimensional analysis as applied above.

APPENDIX 1

The Unit of Measure for Beauty in Geometry

All Greek philosophers agreed that beauty means well-balanced proportions. In architecture, for instance, beauty is achieved through harmonious architectonic shapes, gracefully arranged in space; such forms stir the artistic emotions.

Let us, in this context, consider a segment a of a straight line, divided into two equal parts or into a multitude of non-equal parts b and c. Practice has shown that, among several similar divisions the eye shows a preference for a certain division. Fechner, a well-known psychologist of the nineteenth century, investigating this concept, did the following test : he showed a number of people several different rectangles with sides b and c, and asked them to choose the rectangle they like most. Most chose those rectangles whose sides were in a certain ratio.

Taking the value of the segment a of a straight line and certain singular values for the segments b and c of a straight line, the following proportion can be written :

$$\frac{a}{b} = \frac{b}{c}. \tag{1}$$

As given, this proportion corresponds to a division of the segment a of the straight line according to a mean and to an extreme ratio respectively. The straight-line segment a is split into two straight-line segments b and c, so that (1) can also be written in the form :

$$\frac{b + c}{b} = \frac{b}{c}. \tag{2}$$

Euclid introduced the concept of extreme and mean ratios ; the aim was to mark a clear distinction between these two ratios on one hand and all other ratios which could be written for any other division of a segment of a straight line, on the other.

It is unanimously agreed, even from earliest times, that unequal segments the lengths of which are in an extreme or in a mean ratio to one another convey a feeling of harmony, an agreeable impression of balance. Proportion (1) conveys a feeling of rest, constancy, safety and rhythm. It is said that the Egyptians were well aware of this fact and applied this proportion in the design of the king's room of the Great Pyramid ; indeed, the sizes of this room agree with proportion (1). The ancient Greeks too were aware of this empirical law ; in the sculpture and architecture of Pericles' age, proportion (1) is strictly adhered to. Indeed, this proportion underlies the Greek philosophers' concept of beauty, as stated above.

Fra Luca Pacioli di Borgo (sixteenth century) called relation (1) the divine proportion. This empirical preference of our eyes for a certain ratio between the divisions of a length was why Leonardo Da Vinci called this preferential division "sectio aurea", that is the golden section. To substantiate this name, he took several examples from the proportions of various parts of the human body and argued that its harmony is explained by the fact that the division of the various component parts obeys the golden section rule. This situation is also met with in architecture.

Some 450 years later, Le Corbusier, the famous French architect resurrected the idea of the division according to the golden section ratio and called it in French "le modulor" (*die goldene Schnitt*, in German) the word being derived from "module", i.e. ratio or proportion scale, and "or", i. e. gold.

When a segment a of a straight line is divided into two unequal parts, b and c, and $b > c$, the $\frac{b}{c}$ ratio is called the *golden number*. Matila Ghyka denoted this ratio by the symbol Φ, and hence we can write :

$$\frac{b+c}{b} = \frac{b}{c} = \Phi \qquad (3)$$

from which the following relation follows :

$$1 + \frac{1}{\Phi} = \Phi, \qquad (4)$$

which expresses the fact that the golden number differs by one unit from its inverse. The second-order equation which results from (4) is :

$$\Phi^2 - \Phi - 1 = 0 ; \qquad (5)$$

it admits the positive solution :

$$\Phi = \frac{1 + \sqrt{5}}{2} = 1.61803398875 \ldots.. \qquad (6)$$

On this basis, the mean value admited for the golden number is :

$$\Phi = 1.618. \qquad (7)$$

Hence, it follows that if the length of the straight-line segment is known, and if this segment is divided according to the golden section rule, the length of the shorter segment of the straight line will be :

$$c = \frac{1}{2.618}\, a, \qquad (8)$$

while that of the longer segment will obviously have the value :

$$b = \frac{1.618}{2.618} a. \tag{9}$$

In strict compliance with the empirical observation about the preference of the eye for a certain proportion — which led to the establishment of the golden section and the golden number — we can cite a property of live matter, namely growth resulting from succesive, additional deposits. Thus, botanists found that the length intervals between the knots from which the leaves develop on a plant stem follow a certain law. With animals, the same law is observed in the shape of some mollusc shells, animals' horns or bones — for example the femural bone of human beings ; in this latter case, although it grows from one end only, the original shape is preserved ; it merely grows bigger. These findings in natural sciences led to the formulation of the law of organic growth. The first to draw attention to this law was Fibonacci. He counted the number of pairs of rabbits descending from the same pair, established if they give birth to cubs each month and if, in their turn, the young rabbits were fit to give birth to other cubs, within the same time period of one month. Fibonacci thus arrived at this sequence of numbers :

$$1,\ 1,\ 2,\ 3,\ 5,\ 8,\ 13, \ldots \tag{10}$$

where each number is the sum of the two preceding ones. The general term of sequence (10) has the form :

$$u_n = u_{n-1} + u_{n-2}, \tag{11}$$

where $u_1 = u_2 = 1$, while $n = 3, 4, \ldots$ The sequence of numbers (10) is a double additive sequence and was called the Fibonacci sequence.

Let us consider a Fibonacci sequence of the form :

$$u_1, u_2, \ldots, u_n, \ldots \tag{12}$$

and let us further assume that the general term is the same as in a geometric progression of rate q, that is

$$u_n = q^{n-1}. \tag{13}$$

Based on relation (11), which expresses the characteristic feature of a Fibonacci sequence, we can write :

$$q^{n-1} = q^{n-2} + q^{n-3} ; \tag{14}$$

from which, simplifying by q^{n-3}, we obtain :

$$q^2 - q - 1 = 0. \tag{15}$$

This result is certainly a surprise ; relation (15) has the same form as (5), which is satisfied by the golden number.

Going a step further, let us assume that the golden section is a particular case of the law of organic growths. Now, inasmuch as

$$\frac{4}{\sqrt{\Phi}} = 3.14 \ldots, \tag{16}$$

another known constant will be taken into consideration, namely

$$\pi = 3.14 \ldots \tag{17}$$

APPENDIX 2

Dimensional Equations for Some Physical Quantities in the International System of Units

No. of Order	Designation	Symbol	Definition relation or Definition	Dimensional equation
1	2	3	4	5
	1. Quantities Characteristic of Geometry			
1.1	Distance: length, width, height, thickness, radius	l, x, r	—	L
1.2	Area	A	$A = l_1 l_2$	L^2
1.3	Volume, capacity	V	$V = l_1 l_2 l_3$	L^3
1.4	Plane angle	φ	$\varphi = \frac{(l)}{r}$	—
1.5	Solid angle	Ω	$\Omega = \frac{A}{r^2}$	—
	2. Quantities Characteristic of Kinematics			
2.1	Time	t	—	T
2.2	Velocity	v	$v = \frac{dl}{dt}$	LT^{-1}
2.3	Velocity gradient	grad v	grad $v = \frac{dv}{dx}$	T^{-1}
2.4	Acceleration	ς, g	$\varsigma = \frac{dv}{dt}$	LT^{-2}
2.5	Angular velocity	ω	$\omega = \frac{d\varphi}{dt}$	T^{-1}

	2	3	4	5
2.6	Angular acceleration	$\varkappa$	$\varkappa = \frac{d\omega}{dt}$	T^{-2}
2.7	Period	T	$f(t) = f(t + T)$	T
2.8	Frequency	μ	$\mu = \frac{(1)}{T}$	T^{-1}
2.9	Rotational frequency (number of revolutions)	n	$n = \frac{1}{T}$	T^{-1}
2.10	Angular frequency (pulsation)	Z	$Z = 2\pi\mu$	T^{-1}
2.11	Wavelength	χ	$\chi = l$ between two successive points	L
	3. Quantities Characteristic of Mechanics			
3.1	Mass	m	—	M
3.2	Density	ρ	$\rho = \frac{m}{V}$	$L^{-3}M$
3.3	Density gradient	grad ρ	grad $\rho = \frac{\Delta\rho}{\Delta x}$	$L^{-4}M$
3.4	Mass volume	β	$\beta = \frac{V}{m}$	L^3M^{-1}
3.5	Impulse	I	$I = mv$	LMT^{-1}
3.6	Momentum of a body (angular momentum)	b	$b = Il$	L^2MT^{-1}
3.7	Moment of inertia	j	$j = \int x^2 dm$	L^2M
3.8	Force	F	$F = ma$	LMT^{-2}
3.9	Weight	G	$G = mg$	LMT^{-2}
3.10	Specific weight	γ	$\gamma = \frac{(G)}{V}$	L^2MT^{-2}
3.11	Moment of force	M	$M = Fl$	L^2MT^{-2}
3.12	Force impulse (percussion)	I	$I = \int F\,dt$	LMT^{-1}
3.13	Pressure	p	$p = \frac{F}{A}$	$L^{-1}MT^{-2}$
3.14	Stress	τ	$\tau = \frac{F}{A}$	$L^{-1}MT^{-2}$
3.15	Pressure gradient	grad p	grad $p = \frac{dp}{dx}$	$L^{-2}MT^{-2}$

1	2	3	4	5
3.16	Relative linear expansion	ε	$\varepsilon = \frac{\Delta l}{l_0}$	—
3.17	Elastic modulus	R	$R = \frac{\tau}{\varepsilon}$	$L^{-1}MT^{-2}$
3.18	The second order moment of inertia of the volume	j_v	$j = \int_b x^2 dV$	L^5
3.19	The second order moment of inertia of the area	j_A	$j_A = \int_A x^2 dA$	L^4
3.20	The second order moment of inertia of the line	j_l	$j_l = \int_v x^2 \, dl$	L^3
3.21	Inertia modulus	W	$W = \frac{j_A}{l}$	L^3
3.22	Dynamic viscosity	η	$F = \eta \frac{dv}{dx} A$	$L^{-1}MT^{-1}$
3.23	Kinematic viscosity	ν	$\nu = \frac{\eta}{\rho}$	L^2T^{-1}
3.24	Fluidity	δ	$\delta = \frac{1}{\eta}$	$LM^{-1}T$
3.25	Surface tension	σ	$\sigma = \frac{F}{l}$	MT^{-2}
3.26	Mass flow-rate	Q_m	$Q_m = \frac{m}{t}$	MT^{-1}
3.27	Density of mass flow-rate	q_m	$q = \frac{Q_m}{A}$	$L^{-2}MT^{-1}$
3.28	Volume flow-rate	Q_V	$Q_V = \frac{V}{t}$	L^3T^{-1}
3.29	Density of volume flow-rate	q_V	$q_V = \frac{Q_V}{A}$	LT^{-1}
3.30	Mechanical work	L	$L = Fl$	L^2MT^{-2}
3.31	Mechanical energy	W	$W = Pt$	L^2MT^{-2}
3.32	Impact elasticity (resilience)	K_n	$K_n = \frac{L}{A}$	MT^{-2}

1	2	3	4	5
3.33	Density of mechanical energy	w	$w = \frac{W}{V}$	$L^{-1}MT^{-2}$
3.34	Mass energy	W_m	$W_m = \frac{W}{m}$	L^2T^{-2}
3.35	Mechanical power	P	$P = \frac{w}{t}$	L^2MT^{-3}
3.36	Mass power	P_m	$P_m = \frac{P}{m}$	L^2T^{-3}
	4. Quantities Characteristic of Heat			
4.1	Thermodynamical temperature	θ	—	Θ
4.2	Gradient of thermodynamic temperature	grad θ	$\text{grad}\ \theta = \frac{\Delta\theta}{\Delta x}$	$L^{-1}\ \Theta$
4.3	Coefficient of linear expansion	d_1	$d_1 = \frac{1}{l}\left(\frac{dl}{d\theta}\right)$	Θ^{-1}
4.4	Coefficient of volumic expansion	d_V	$d_V = \frac{1}{V}\left(\frac{dV}{d\theta}\right)$	Θ^{-1}
4.5	Coefficient of pressure build-up	d_p	$d_p = \frac{1}{p}\left(\frac{dp}{d\theta}\right)$	Θ^{-1}
4.6	Coefficient of isothermal compression	ξ	$\xi = \frac{1}{V}\left(\frac{dV}{dp}\right)$	$LM^{-1}T^2$
4.7	Heat amount	Q_c	$Q_c = W$ transferred with no mechanical work done	L^2MT^{-2}
4.8	Heat flux	Φ	$\Phi = \frac{dQ_c}{dt}$	L^2MT^{-3}
4.9	Density of heat flux	q	$q = \frac{d\Phi}{dA}$	MT^{-3}
4.10	Thermal conductivity	λ	$q = \lambda\ \text{grad}\ \theta$	$LMT^{-3}\Theta^{-1}$

1	2	3	4	5
4.11	Coefficient of heat transfer	α	$q = \alpha\Delta\theta$	$MT^{-3}\Theta^{-1}$
4.12	Thermal diffusivity	a	$a = \frac{\lambda}{\rho C}$	L^2T^{-1}
4.13	Heat capacity	C	$C = \frac{dQ_c}{d\theta}$	$L^2MT^{-2}\Theta^{-1}$
4.14	Mass heat	c	$c = \frac{dQ_c}{m d\theta}$	$L^2T^{-2}\Theta^{-1}$
4.15	Latent heat	r	$r = Q_c$ taken or given up through an isothermal conversion	L^2MT^{-2}
4.16	Mass latent heat	r_m	$r_m = \frac{r}{m}$	L^2T^{-2}
4.17	Mass latent heat for expansion	r_d	$r_d = \frac{1}{m}\left(\frac{\partial Q}{\partial V}\right)$	$L^{-1}T^{-2}$
4.18	Mass latent heat for compression	r_c	$r_c = \frac{1}{m}\left(\frac{\partial Q}{\partial p}\right)$	L^3M^{-1}
4.19	Transitional heat	Q_t	$Q_t = Q_c$ for the transition from a molecular state into another molecular state	L^2MT^{-2}
4.20	Mass transitional heat	q_t	$q_t = \frac{Q_t}{m}$	L^2T^{-2}
4.21	Entropy	S	$dS = \frac{Q_t}{\theta}$	$L^2MT^{-2}\Theta^{-1}$
4.22	Mass entropy	s	$s = \frac{S}{m}$	$L^2T^{-2}\Theta^{-1}$
4.23	Internal energy	U	$dU = dQ_c + AdL$	L^2MT^{-2}
4.24	Heat content (enthalpy)	H	$H = U + pV$	L^2MT^{-2}
4.25	Free energy	W_l	$W_l = U - \theta S$	L^2MT^{-2}
4.26	Free heat content (enthalpy)	H_l	$H_l = H + \theta S$	L^2MT^{-2}

1	2	3	4	5
4.27	Internal mass energy	u	$u = \frac{U}{m}$	L^2T^{-2}
4.28	Mass enthalpy (heat content)	h	$h = \frac{H}{m}$	L^2T^{-2}
4.29	Mass free energy	w_l	$w_l = \frac{W_l}{m}$	L^2T^{-2}
4.30	Mass free heat content (enthalpy)	h_l	$h_l = \frac{H_l}{m}$	L^2T^{-2}
	5. Quantities Characteristic of Molecular Physics			
5.1	Concentration	ρ_B	$\rho_B = \frac{m_B}{V_B}$	L^3M
5.2	Diffusion flux	ψ	$\psi = \frac{\mathrm{d}m}{\mathrm{d}t}$	MT^{-1}
5.3	Density of diffusion current	i	$i = \frac{\mathrm{d}\psi}{\mathrm{d}A}$	$L^{-2}MT^{-1}$
5.4	Diffusion coefficient	D	$i = -D \operatorname{grad} \rho$	L^2T^{-1}
5.5	Molar fraction of the B component	x_B	The ratio between the amount of substance of component B and that of the mixture	—
5.6	Ratio of thermal diffusion	k_θ	$\operatorname{grad} x_B = \frac{k_\theta}{\theta} \operatorname{grad} \theta$	—
5.7	Coefficient of thermal diffusion	D_θ	$D_\theta = k_\theta D$	L^2T^{-1}

APPENDIX 3

The Unit: in Physics, Chemistry and Nuclear Physics

Physics is concerned with the study of physical phenomena, their states and transformations. To describe any well-determined physical phenomenon, a finite, comparatively small number of variables are used.

All dimensional variables have dimensional equations of the form

$$\mathrm{L}^{la}\,\mathrm{M}^{ma}\mathrm{T}^{ta}\ldots \tag{1}$$

Between the variables which describe a physical phenomenon, a certain relation exists. In terms of general dimensional analysis, this relation is written

$$M^m N^n P^p \ldots = kA^a B^b C^c \ldots \tag{2}$$

Physical relations should be dimensionally homogeneous. The homogeneity condition is expressed by systems of equations of the form

$$\Sigma v_1 v_2 = 0, \tag{3}$$

where v_1 and v_2 stand for the dimensional and quantity exponents respectively. These systems of equations are indeterminate.

In general dimensional analysis, the variables are distributed before writing the homogeneity condition ; in this way, the negative solutions of the systems of equations are eliminated and the equations are solved only for integral solutions.

Chemistry is concerned with the study of chemical substances and their transformations. Chemical substances are combinations of chemical elements and are expressed by chemical formulae. Transformations of chemical substances are called chemical reactions. During a chemical reaction, one or several chemical substances are transformed into one or several new chemical substances. In any chemical reaction, the total mass of combining substances is maintained constant; the variable is the mass of molecular species : nothing is lost, nothing is gained and all is transformed. This is the mass conservation law. In the general case of a chemical reaction, the participants are the initial substances and the resultant substances.

Each of the substances represent combinations of chemical elements ; so let us write

$$A = X_{x_a} Y_{y_a} Z_{z_a} \dots \tag{4}$$

In the equations written for the substances, elemental coefficients $x_a, y_a, z_a, \dots$ appear and these take integral positive values.

The chemical equation is written in the form

$$aA + bB + cC + \dots = mM + nN + pP + \dots ; \tag{5}$$

substance coefficients are ascribed to the substances thus : $a, b, c, \dots$ The mass conservation condition is expressed by the systems of equations, still in the form

$$\Sigma v_1 v_2 = 0. \tag{3}$$

The initial substances, as well as those resulting from any reaction, are distributed from the very beginning between the two members of a chemical equation. In such a situation, the unknowns will admit only positive values.

The molecule of any chemical substance always contains an integral number of atoms ; this fact is born in mind when establishing the chemical formulae. The substance coefficients should also take only integral values, since chemical reactions only admit integral numbers of atoms and molecules.

As a general rule the formulae set up for chemical substances will contain the least coefficients of chemical elements, for the sake of simplicity. The same concern for simplicity dictates that the chemical equations have substance coefficients with minimum values.

Nuclear physics is concerned with the atomic nucleus and its structural changes. Any such change is a nuclear reaction. In nuclear reactions elemental particles also appear ; or it may be that as a consequence of a nuclear reaction energy is developed in the form of radiations. Nuclear reactions satisfy the law of energy conservation. In the general case of a nuclear reaction, initially only a nucleus ${}^{A}_{Z}X$ and an elemental particle ${}^{A_1}_{Z_1}Y_1$ are present, while the reaction result is a nucleus ${}^{A'}_{Z'}X$ and either another nucleus ${}^{A''}_{Z''}X$ or an elemental particle ${}^{A'}_{Z'}Y$, such as, for instance, in the reaction :

$$f({}^{A}_{Z}X,\ {}^{A_1}_{Z_1}Y,\ {}^{A'}_{Z'}X,\ {}^{A''}_{Z''}X) = 0. \tag{6}$$

Hence, for nuclei and elemental particles we can write :

$$\left.\begin{aligned} {}^{A}_{Z}X &= f(A, Z) \\ {}^{A_1}_{Z_1}Y &= f(A_1, Z_1), \end{aligned}\right\} \tag{7}$$

where A is the mass number and Z is the atomic number.

The most general form of a nuclear reaction is represented, say, by the equation

$$a_Z^A X + b_{Z_1}^{A_1} Y \rightarrow c_{Z'}^{A'} X' + d_{Z''}^{A''} X'' \qquad (8)$$

where $a, b, c \ldots$ are either nucleus coefficients or elemental particle coefficients.

Equation (8) is valid only when the conditions required by the conservation law of the number of nucleons and of electric loads are simultaneously adhered to; this conservation law is expressed by systems of equations of the same form as above, i.e.

$$\Sigma v_1 v_2 = 0. \qquad (3)$$

The initial nuclei and elemental particles, as well as the resulting ones, will be written from the start in the left-hand and in the right-hand members respectively, of a nuclear equation. As a consequence, the unknowns will admit only positive values.

The nuclei always contain integral numbers of nucleons, i.e. neutrons and protons. As a consequence, nucleus and elemental particle coefficients should also take only integral values since in nuclear reactions only integral numbers of nuclei and elemental particles can take place.

The simplest form of a nuclear equation is achieved by adopting minimum values for the nucleus and for the elementary particle coefficients.

The systems of equations (3) are Diophantine and express the homogeneity condition in physics, the mass conservation condition in chemistry and the condition of nucleon and electric load conservation in nuclear physics.

The identity of systems of equations (3) underlines the dialectic unity of these three fields of science.

APPENDIX 4

An Algorithm Used with the Method of General Dimensional Analysis

To describe complex physical processes and phenomena, a greater number of variables are needed ; in such cases, the bulk of calculations also considerably increases and therefore one has to have recourse to mechanized means of computation. In what follows, we shall present some elementary knowledge of how the algorithm used to derive the physical relations from the dimensional equations and from the values of measurable quantities is established[1].

All the quantities used in physics or in other fields to which the physical phenomena of the processes to be studied by dimensional analysis belong are entered in the computer store. The computer's store should also contain the dimensional equations of all these quantities, which may be either primary or secondary, and should be written in the International System of Units ; however, for the calculations required in general dimensional analysis, any system of units is acceptable.

The user should specify from the beginning the list of quantities needed for the investigation of a physical phenomenon or process ; for such quantities, he shall also enter into the computer the sets of the numerical values he intends to use in deriving the physical relation looked for.

For the algorithm described in what follows, "structure" will denote that form of a physical relation of the monomial type where the quantities are distributed between the two members. If, subsequently, all quantities are written in one member, their distribution will be reflected by the signs ("+" or "−") ascribed to the values of the quantity exponents ; these signs will show the correlation sense of variables in each case under study. Changing the signs of the quantity exponents in a given relation leads to a new structure of the said relation. In general dimensional analysis, before starting the cal-

---

[1] Constantin, Ştefana, *An algorithm to describe a physical phenomenon through general dimensional analysis*, Report presented to the Informatics and Management Symposium, Cluj—Napoca, May, 1980.

culations, unknown exponents will be ascribed to all measurable quantities included in the relation to be derived.

The dimensional equation of a physical quantity is expressed by a vector thus specified :

— it shall include the same number of components as the fundamental quantities established for the relevant field of physics or technical science ; these quantities shall be considered in the order chosen by the user ;

— the value of each component represents the dimensional exponent, that is the exponent of the fundamental quantity in the dimensional equation of the relevant physical quantity.

Having these vectors as columns, a dimensional matrix of $d^* \times m^*$ elements is built up, where d^* is the number of fundamental quantities while m^* is the number of physical quantities describing the phenomenon under study. Since, in the calculations implied by general dimensional analysis, only non-zero values effectively occur, these should be extracted from the matrix in view of their use and also to avoid possible errors. These operations are carried out by the VALEF sub-routine (VALues ; EFfective).

When even one single line of the dimensional matrix contains one non-zero element, the list of quantities is not correct and the calculations will be interrupted.

The algorithm should be able to build up the structure of the computing relatioe by itself. The change-over from one structure to another with a view to obtaining th correct structure is performed by the SIGNUM sub-routine.

In the system of linear equations expressing the condition of dimensional homogeneity for the computing relation, the coefficients are the values of dimensional vectors while the unknowns are the quantity exponents ; the plus and minus signs account for the relation structure at the relevant stage of the computations. The equations contain no non-zero terms. The system of equations is solved by iteration, in the increasing order of terms ; the solution should be chosen with due consideration paid to the theorem of positive, integral, minimum values. These computations are carried out by sub-routine SOLEQ (SOLution ; EQuation).

The numerical values of physical quantities are used to compute the value of the non-dimensional numerical factor k, by means of sub-routine DETK (DETermination of k). For each set of data, a value for the non-dimensional numerical factor will result. The standard deviation σ of the values obtained is estimated ; it is considered that the numerical factor has a constant value when $\sigma \leqslant 10^{-n}$; the value of n is chosen by the user. Now, the values of the quantity exponents are correct.

Should the standard deviation fail to satisfy the condition above, the computations shall be resumed by : changing the relation structure ; choosing a new solution for the new system of equations ; computing the values of the non-dimensional numerical factor with the same experimental data ; and checking the condition for the value of the standard deviation. When all possible combinations allowed by the physical quantities included in the list originally adopted are tried out and no practically constant value is obtained for the non-dimensional numerical factor, the physical phenomenon under study should be reconsidered, or else, small, positive, integral values shall be adopted for the quantity exponents.

The block-diagram of the computing algorithm is shown in the figures on pages 291—2.

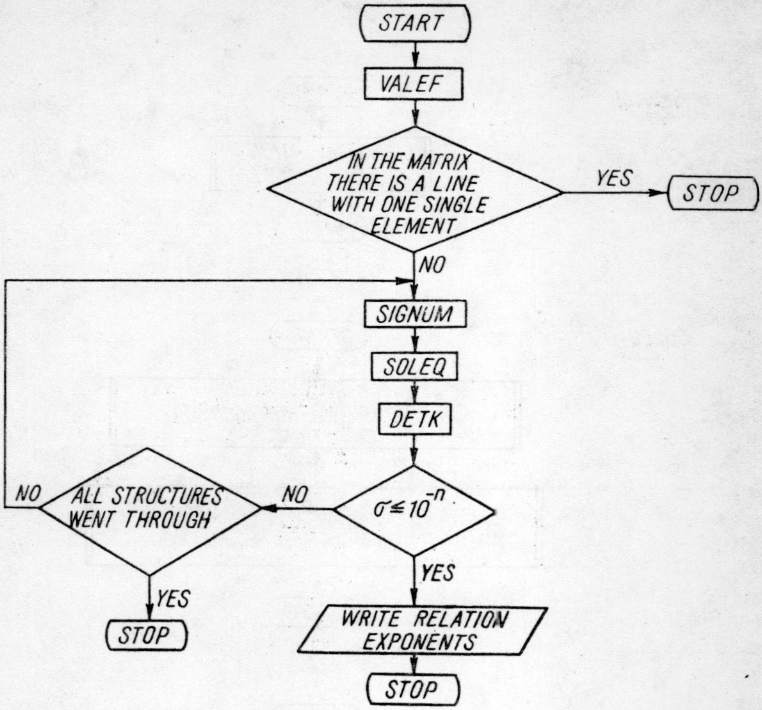
START
VALEF
IN THE MATRIX THERE IS A LINE WITH ONE SINGLE ELEMENT
YES
STOP
NO
SIGNUM
SOLEQ
DETK
$\sigma \leq 10^{-n}$
NO
ALL STRUCTURES WENT THROUGH
NO
YES
STOP
YES
WRITE RELATION EXPONENTS
STOP

VALEF:

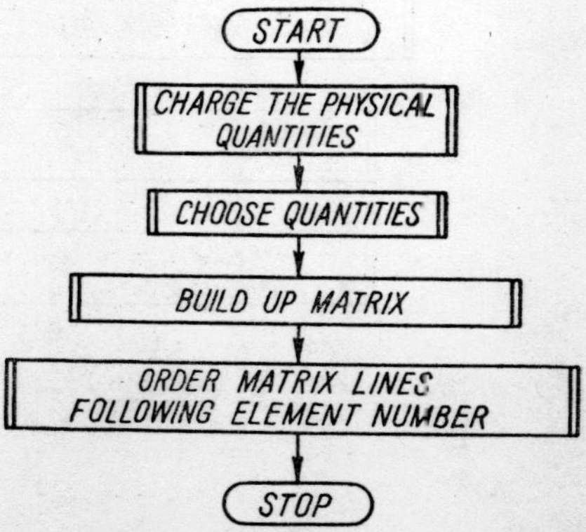
START
CHARGE THE PHYSICAL QUANTITIES
CHOOSE QUANTITIES
BUILD UP MATRIX
ORDER MATRIX LINES FOLLOWING ELEMENT NUMBER
STOP

SIGNUM:

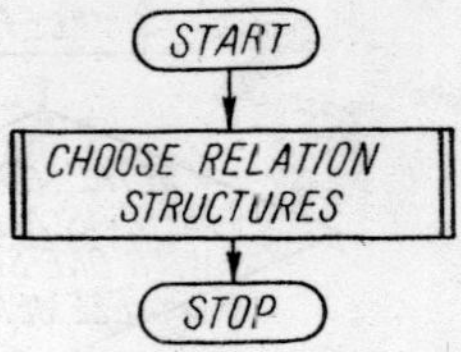

SOLEQ:

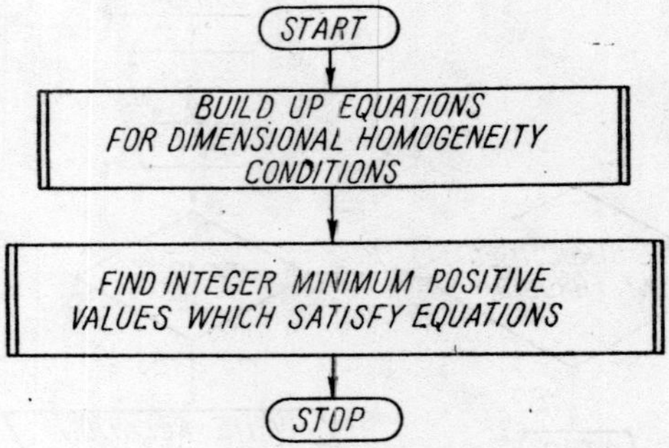

DETK:

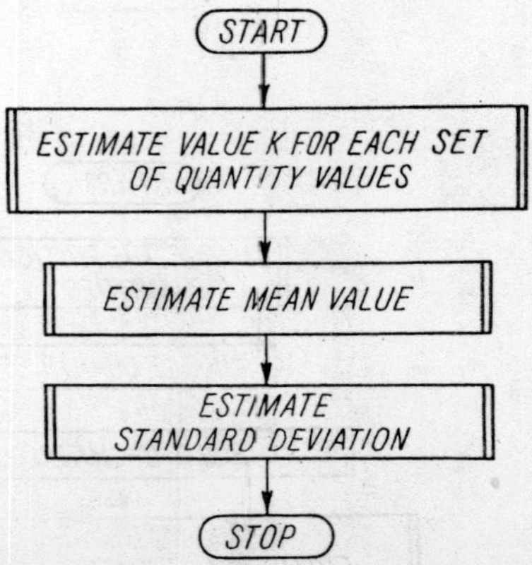

List of references

Chapter I: The Basis of Dimensional Analysis

1.1. APOSTOL P., *Experiment and Knowledge*, Bucharest, Editura Ştiinţifică, 1954.

1.2. ARNOLD, R. R., HILL, H. C., NICHOLS, A. V. *An Introduction to Data Processing*, New York, John Wiley & Sons, 1966.

1.3. BAAR, D. I. H., The proportionalities method of dimensional analysis, *Journal of the Franklin Institute*, CCLXXXXII, No. 6, Dec. 1971, pp. 441–449.

1.4. BĂRBULESCU, N., *Physics*, Vol. I, Bucharest, Editura Didactică şi Pedagogică, 1972.

1.5. BELLA, A., *Ship's Dynamics* (Dinamica de la Nave), Vol. I, Genova, Editore Briano, 1951.

1.6. BRIDGMAN, P. W., *Dimensional Analysis*, New Haven, Yale University Press, 1943.

1.7. CHEN, WAI-KAI, Algebraic theory of dimensional analysis, *Journal of the Franklin Institute*, CCLXXXXII, No. 6, Dec. 1971, pp. 403–422.

1.8. COMOLET, R., *Introduction à l'Analyse Dimensionnelle et aux Problèmes de Similitude en Mécanique des Fluides*, Paris, Masson, 1958.

1.9. DOUGLAS, J. F., *An Introduction to Dimensional Analysis for Engineers*, London, Pitman Press, 1969.

1.10. DUNCAN, W. I., *Physical Similarity and Dimensional Analysis. An Elementary Treatise*, London, Edward Arnold, 1953.

1.11. ESNAULT-PELTERIE, R., *L'analyse Dimensionnelle*, Lausanne, Editions F. Rouge, 1946.

1.12. ESNAULT-PELTERIE, R., *Analyse Dimensionnelle et Métrologie*, Lausanne, Editions F. Rouge, 1950.

1.13. ESNAULT-PELTERIE, R., *Analyse Dimensionnelle et Métrologie*, Bruxelles, Editions techniques et scientifiques R. Louis, 1952.

1.14. FLORESCU, M., *The Matter or the Objective Reality*, Bucharest, Editura Politică, 1972.

1.15. FLORESCU, M., *Dimensions of Knowledge*, Bucharest, Editura Politică, 1977.
1.16. FULLER, A. T., Dimensional properties of optimal and suboptimal non-linear control systems, *Journal of the Franklin Institute*, CCLXXXIX, No. 5, Nov., 1970, pp. 379–393.
1.17. GÖRTLER, H., *Dimensional Analysis. Theorie de: physikalishen Dimensionen mit Anwendungen*, Berlin, Springer, 1977.
1.18. HAMBURGER, L., *Introduction to the Theory of Heat Propagation*, Editura Academiei R.P.R., 1951.
1.19. HUNTLEY, H. E., *Dimensional Analysis*, New York, Dover Publications, Inc., 1967.
1.20. ISCRULESCU, I., ISPĂȘOIU, GH., and PETRESCU V., *The International System of Units*, București, Editura Tehnică, 1970.
1.21. JOLTEA, A., The Phylosophic Substrate of the Continuum Hypothesis and Rationality of Knowledge, *Forum IV*, No. 4, 1972, pp. 111–135.
1.22. KIRPICEV, M. V., Analiz razmernosti, *Izvestia Academii Nauk, Sov. Union—Otdelenie Techniceskih Nauk*, No. 9, 1953, pp. 1330–1340.
1.23. KOLONITS, F., Hosonloságelméletes dimenzióanalizis, *Energia ás Atomtechnica*, XXIV, No. 9, Sept. 1971, pp. 353–357.
1.24. KRON, R. P., Dimensions, *Journal of the Franklin Institute*, CCLXXXXII, No. 1, July 1971, pp. 45–55.
1.25. LANGHAAR, H. L., A Summary of dimensional analysis, *Journal of the Franklin Institute*, CCXXXXII, No. 6, Dec. 1946, pp. 459–463.
1.26. LANGHAAR, H. L., *Dimensional Analysis and Theory of Models*, New York, John Wiley & Sons, 1951.
1.27. MACAGNO, E. O., Historico-critical review of dimensional analysis, *Journal of the Franklin Institute*, CCLXXXXII, No. 6, Dec. 1971, pp. 391–402.
1.28. MARTINOT-LAGARDE, A., *Analyse Dimensionnelle. Application à la Mécanique des Fluides*, Paris, Office National d'Etudes et des Recherches, 1948.
1.29. MELDHAL, A., Uber homologen geschribene Gleichungen, Massysteme und Dimensionanalyse, *BBC Nachrichten*, No. 6, 1961, pp. 319–323.
1.30. MORAN, M. J., A generalisation of dimensional analysis, *Journal of the Franklin Institute*, CCLXXXXII, No. 6, Dec. 1971, pp. 423–432.
1.31. ONCESCU, M., *Quantities and Units in Physics*, Vol I, Bucharest, Editura Tehnică, 1955.
1.32. PALACIOS, J., *Analyse Dimensionnelle* (translated from Spanish), Paris, Gauthier Villars, 1960.
1.33. RĂDULEȚ, R., *Theoretic Basis of Electrical Engineering*, Bucharest, Editura Institutului Politehnic, 1953.
1.34. RUMȘINSCHI, L. Z., *Mathematic Processing of Experimental Data* (translated from Russian), Bucharest, Editura Tehnică, 1974.
1.35. SAINT-GUILHEM, L. Z., Système des unités et l'analyse dimensionnelle, *Annales des Mines*, CXXXVII, No. 5, 1948, pp. 5–41.
1.36. SEDOV, L. I., *Metodi Podobija i Razmernosti v Mehanike*, Moscow, Izdatelstvo Nauka, 1965.
1.37. TIRON, M., *Statistic and Informational Processing of Measurement Data*, Bucharest, Editura Tehnică, 1976.
1.38. TODORAN, I., *Mathematical Processing of Experimental Data*, Bucharest, Editura Academiei R.S.R., 1976.
1.39. ȚIȚEICA, R. and POPESCU, I., *General Physics*, Bucharest, Editura Tehnică, 1971.

1.40. VALLOT, J., *Grossengleichungen Einleiten und Dimensionen*, Leipzig, Johann Ambrosius Barth, 1957.
1.41. VASILESCU, A. A., *Dimensional Analysis and the Theory of Similitude*, Editura Academiei R.S.R., 1969.
1.42. WORTHING, A. G., GEFNER, I., *Treatment of Experimental Data*, New York, John Wiley & Sons, London, Chapman & Hall, 1955.
1.43. WALD, H., *Elements of General Epistemology*, Bucharest, Editura Ştiinţifică, 1967.
1.44. *The Romanian Technical Lexicon, Vol. I—VII*, Bucharest, Editura Tehnică, 1949—1955.
1.45. Modern dimensional analysis, similitude and similarity; Special issue: *Journal of the Franklin Institute*, CCLXXXXII, No. 6, Dec. 1971.
1.46. *Dictionary of Physics*, Bucharest, Editura Enciclopedică română, 1972.
1.47. *Dictionary of General Mathematics*, Bucharest, Editura Enciclopedică Română, 1974.

Chapter II: Restricted Dimensional Analysis

2.1. ADAMS, W. H. MC., *Heat Transmission*, New York, Toronto, London, McGraw-Hill Book Co. Inc., 1942.
2.2. ALABUJEV, P. M., GHERONIMUS, V. B., MINKEVICI, L. M. and SEHOBTOV B. A., *Teorii Podobija i Razmernostei Modelirovanie*, Moscow, Izdatelstvo Vîsshaïa Skola, 1969.
2.3. BIRKHOFF, G., *Hydrodynamics, a Study in Logic, Fact and Similitude*, Cincinnati, Princeton University Press, 1950.
2.4. BRIAN, E., *Basic Concepts of Measurement*, London, Cambridge University Press, 1969.
2.5. BRUN, E. A. and MARTINOT-LAGARDE, A., *Mécanique des Fluides*, Vol. I, Paris, Dunod, 1968.
2.6. CIOC D., *Fluid Mechanics*, Bucharest, Editura Didactică şi Pedagogică, 1967.
2.7. COMOLET, R., *Mécanique Expérimentale des Fluides*, Vol. I & II, Paris, Masson, 1961, 1963.
2.8. GALBURĂ, GH., *Introduction to Algebra*, Bucharest, Editura Tehnică, 1969.
2.9. GERMANI, D., *Sur la Structure des Formules et la Synthèse des Lois de Similitude en Physique. Considérations sur les Grandeurs et Leurs Unités*, Bucharest, Imprim. Naţională, 1931.
2.10. GRIGOROVICI, R. and ONCESCU, M., *Quantities and Units in Physics*, Vol. II, Bucharest, Editura Tehnică, 1958.
2.11. HAINZL, J., On local generalizations of the Pi-theorem of dimensional analysis, *Journal of the Franklin Institute*, CCLXXXXII, No. 6, Dec. 1971, pp. 463—470.
2.12. IOACHIMESCU, A. G., *Rational Mechanics*, Bucharest, Imprimeria Naţională, 1947.
2.13. IVERNEL A., *Représentations Analogiques et Homologiques dans les Techniques de la Chaleur*, Paris, Dunod, 1965.
2.14. IACOVLEFF, A., *Eléments de la Théorie des Moteurs à Combustion Interne à Régime Rapide*, Paris, Eyrolles Editeur, 1948.
2.15. KAY, M. T., *Introduction à la Mécanique des Fluides et la Transmission de la Chaleur* (translated from English), Paris, Dunod, 1964.
2.16. KOLONTIS, F., Teljes Készeletek höátadási viszgálatokhoz, *Energia és Atomtechnika*, XXIV, No. 12, Dec. 1971, pp. 506—515.

2.17. MARTINOT-LAGARDE, A., *Essai Synthétique sur la Similitude Physique*, Paris, Conférence à l'École Centrale des Arts et Manufactures, 1971.

2.18. MATEESCU, C., *Hydraulics*, Bucharest, Editura Didactică și Pedagogică, 1963.

2.19. MATZ, W., *Le principe de Similitude en Génie Chimique*, Paris, Dunod, 1959.

2.20. OUZIAUX, R. and PERRIOR, I., *Mécanique des Fluides Appliquée*, Vol. I, Paris, Dunod, 1966.

2.21. PANKHURST, R. C., Alternative formulation of the Pi-theorem, *Journal of the Franklin Institute*, CCLXXXXII, No. 6, Dec. 1971, pp. 451–462.

2.22. PAWLOWSKI, J., *Die Aehnlichkeits-theorie in der Technish-Physikalishen Forschung*, Berlin, Julius Springer, 1971.

2.23 PLĂCINȚEANU, I. I., Physical Quantities and their Units, *Tehnica Militară* Supplement, No. 1.

2.24. PROCOPIU, S., *Electricity and Magnetism*, Iași, Editura Laboratorului de Electricitate, 1942.

2.25. RĂDULEȚ, R., Les espèces de grandeurs électromagnétiques primitives; *Revue d'Electrotechnique et Energétique*, *A.*, No. 1, 1963, pp. 1–27.

2.26. STOENESCU, A., *Units of Measurement in Mechanics*, Bucharest, Editura Institutului Politehnic, 1951.

2.27. VASCHY, A., Sur les lois de similitude en physique; *La Houille Blanche*, XXV, No. 1, 1970, pp. 72–74.

2.28. VENIKOV, V. A., *Application of Similitude Theory and of Simulation to Electromagnetics* (translated from Russian and completed), Bucharest, Editura Tehnică, 1972.

2.29. *A Dictionary of Technical Terms*, Bucharest, Editura Tehnică, 1973.

Chapter III: Applications of the Restricted Dimensional Analysis

3.1. ARJANIKOV, N. S. and MALTZEV, V. N., *Aerodynamics* (translated from Russian), Bucharest, Editura tehnică, 1954.

3.2. BRUHAT, G., *Thermodynamique*, Paris, Masson, 1939.

3.3. CHAKRABARTI, S. K., On constitutive equations and dimensional analysis, *Journal of the Franklin Institute*, CCLXXXXVI, No. 5, Nov. 1973, pp. 339–346.

3.4. DRAGOȘ, L., Gnoseologic aspects of principles of mechanics, *FORUM*, IV, No. 4, 1972, pp. 89–97.

3.5. DRĂGHICI, N. N., *Pipes for Fluids Transport*, Bucharest, Editura Tehnică, 1971.

3.6. FOKIN, K. F., *Stroitelnaia Teplotechnika Ograjdainisch Ciastei Zdanii*, Moscow, Gosstroizdat, 1953.

3.7. GHIA, V. V. and JIANU, C., Estimation of the flow-rate coefficient of swirl-chamber, frontal return injectors, *Studii și Cercetări de energetică și electrotehnică*, XIX, No. 1, 1969, pp. 3–12.

3.8. LANDAU, L. and LIFCHITZ, E., *Mécanique des Fluides*, Moscow, MIR, 1970.

3.9. LANDSBERG, S. G., *Optics* (translated from Russian), Bucharest, Editura Tehnică, 1951.

3.10. LĂBUȘCĂ, E., Contributions to the study of phenomena taking place during metallic powder pressing, *Studii și Cercetări de Fizică*, V, No. 3—4, 1954, pp. 225—236.
3.11. MIHEEV, M. A., *Basic Heat Transfer* (translated from Russian), Bucharest, Editura Energetică de Stat, 1953.
3.12. MOLINIER, J., ANGELINO, H., Influence de la pulsation sur le transfert de la matière à l'intérieur d'une goutte, *Revue Générale de Thermique*, XVI, No. 191, 1977, pp. 783—793.
3.13. REICHE, F., *Die Quantentheorie*, Berlin, Julius Springer, 1921.
3.14. SPOLSKI, E. V., *Nuclear Physics* (translated from Russian), Vol. I, Bucharest, Editura Tehnică, 1952.
3.15. STAICU, C. I., An equation of invariants characteristic of solidification and melting, *Studii și Cercetări in Energetică*, II, No. 1, 1972, pp. 3—8.
3.16. SIARA, V., Effects of outdoor temperature in building design, *Standardizarea*, VI, No. 5, 1954, pp. 1—8.
3.17. TEODORESCU, C. C., *Strength of Materials*, Bucharest, Imprimeria C.F.R., 1945.
3.18. ȚIȚEICA, S., *A Course on Statistical Physics and Quantum Mechanics*, Bucharest Editura Universității, 1951.
3.19. VASILESCU, A. A., *A Course on Hydromechanics*, Bucharest, Editura Didactică și Pedagogică, 1962.

Chapter IV: General Dimensional Analysis

4.1. GALBURĂ, GH., *Introduction to Algebra*, Bucharest, Editura Tehnică, 1969.
4.2. GHELFOND, A. O., *Resolution of Equations in Integer Numbers* (translated from Russian), Bucharest, Editura Tehnică, 1954.
4.3. LAURENT, M. N., Sur le calcul des dérivées à indices quelconques, *Nouvelles Annales de Mathématiques*, III, No. 3, 1885, pp. 240—252.
4.4. LIOUVILLE, J., Sur le calcul des différentielles à indices quelconques, *Journal de l'Ecole Polytechnique*, XIII, No. 21, 1832, pp. 71—162.
4.5. NOVOSELOV, S. I., *A Special Course on Elemental Algebra* (translated from Russian), Bucharest, Editura Tehnică, 1955.
4.6. STAICU, C. I., Dimensional analysis. Deriving the monomial and homogeneous relations, *Studii și Cercetări de Fizică*, V, No. 3—4, 1954, pp. 237—248.
4.7. STAICU, C. I., Relations derived through dimensional analysis, *Metrologia Aplicată*, III, No. 4, 1956, pp. 25—31.
4.8. STAICU, C. I., On the establishement of new physico-chemical relations, *Studii și Cercetări de Fizică*, IX, No. 4, 1958, pp. 562—567.
4.9. STAICU, C. I., Applications of homogeneity principles to conservation. Deriving the physico-chemical relations on the basis of the principle of integral, small, positive values, *Metrologia Aplicată*, VI, No. 1, 1959, pp. 24—28.
4.10. STAICU, C. I., Fundamental theorem of general dimensional analysis, *Studii și Cercetări Științifice — Fizică și Științe Tehnice*, XIII, No. 2, 1962, pp. 159—172.

4.11. STAICU, C. I., The Law of integral, minimal positive values in general dimensional analysis, *Buletinul Ştiinţific şi Tehnic al Institutului Politehnic Timişoara*, XII (XXVI), No. 6, 1967, pp. 393–401.

4.12. STAICU, C. I., General dimensional analysis, *Buletinul Institutului Politehnic Bucharest*, XXX, No. 2, 1968, pp. 31–44.

4.13. STAICU, C. I., The Law of progressive homogenization in general dimensional analysis, *Buletinul Ştiinţific şi Tehnic al Institutului Politehnic Timişoara*, XIV, (XXVIII), No. 2, 1969, pp. 496–478.

4.14. STAICU, C. I., Nouvelle théorie basée sur l'expérimentation du transfert de la chaleur, Communication présentée au 4e Congrès International sur le transfert de la chaleur, Paris, Versailles, Sept., 1970; published in: *Heat Transfer*, 1970, Amsterdam, Elsevier Publishing Co., 1971.

4.15. STAICU, C. I., Contribution au développement de l'analyse dimensionnelle générale, *Buletinul Inst. Politehnic Iaşi*, XVI (XX), No. 3–4, 1970, pp. 109–116.

4.16. STAICU, C. I., General dimensional analysis, *Journal of the Franklin Institute*, CCLXXXXII, No. 6, 1971, pp. 433–439.

4.17. STAICU, C. I., Sur la forme monome d'une relation physique rationnelle. Communication presentée à la: First Heat Transfer Conference, Iaşi, Oct. 1973; published in: *Preprints of papers presented at the First Heat Transfer Conference*, Iaşi, 1973.

4.18. STAICU, C. I., Against the coefficient for the heat exchange through convection, *Energetica*, XXV, No. 1, 1977, pp. 37–39.

4.19. STAICU, C. I., Experimenting with and processing of measurement data through general dimensional analysis, *Energetica, A Supplement for a Unitary Guidance in Electric Power*, V, 1977, pp. 13–20.

Chapter V: Applications of the General Dimensional Analysi

5.1. CURIEVICI, I., PERETZ, D., PERLMAN, M., SEGAL, A., Some considerations on the inadequateness of the classic concept of equivalent diameter in the numerical analysis of heat flow through elliptic pipes, *Buletinul Institutului Politehnic din Iaşi*, XVIII (XXII), No. 1–2, 1972, pp. 59–64.

5.2. FERRARI, F., Théorie du puits en écoulement non permanent, *La houille blanche* XIII, No. 8, 1958, pp. 784–797.

5.3. GROSU, D., *Planck's System of natural units and the determination of Blackett's formula, Adamachi*, XXXV, Nr. 3–4, 1949.

5.4. JACOVLEFF, D., *Eléments de la Théorie des Moteurs à Combustion Interne à Régime Rapide*, Paris–Liège, Eyrolles–Dresser, 1948.

5.5. KLARSFELD, S. M., Champs de température associés aux mouvements de convection naturelle dans un milieu poreux limité, *Revue Générale de Thermique*, IX, No. 108, 1970, pp. 1403–1423.

5.6. POPOV, M., Fluid spray tests on models, *Studii şi Cercetări de Mecanică Aplicată*, VII, No. 1, 1956, pp. 9–27.

5.7. PROCOPIU, S., Enigma of earth magnetism, *Adamachi*, XXXIV, No. 1–2, 1948.

5.8. RUCKENSTEIN, E., On heat transfer caused by boiling, *Studii şi Cercetări de Energetică*, VIII, No. 1, 1958.

5.9. SCOVLEV, A. J., Heat transfer in the case of thermal periodic phenomena (translated from Russian), Bucharest, *Institutul de Documentare Tehnică*.

5.10. STAICU, C. I., Deriving Blackett's relation through dimensional analysis, *Gazeta Matematică și Fizică*, A, VII (LX), No. 2, 1955, pp. 85–87.

5.11. STAICU, C. I., Stationary convection in free flow, *Studii și Cercetări de Energetică*, V, No. 3–4, 1955, pp. 317–329.

5.12. STAICU, C. I., Determination of the velocity of fall of a sphere through a viscous medium, by means of dimensional analysis, *Gazeta Matematică și Fizică, A*, VIII (LXI) No. 11, 1956, pp. 583–585.

5.13. STAICU, C. I., Deriving some relations in the theory of hydraulic machines, through dimensional analysis, *Energetica*, IV, No. 1956, 12, pp. 580–582.

5.14. STAICU, C. I., On the stationary convection in forced, turbulent air flow, *Studii și Cercetări de Energetică*, VI, No. 4, 1956, pp. 391–401.

5.15. STAICU, C. I., Application de l'analyse dimensionnelle à l'étude de la convection de la chaleur, *Revue d'Electrotechnique et d'Energétique*, XI, No. 1, 1957, pp. 147–162.

5.16. STAICU, C. I., The study of thermic relation, *Studii și Cercetări de Energetică*, VII, No. 3, 1957, pp. 65–86.

5.17. STAICU, C. I., Determination of specific speed of hydraulic turbines through dimensional analysis, *Energetica* V, No. 10, 1957, pp. 505–509.

5.18. STAICU, C. I. and BELDIMAN, M. T., Correction of minimum conventional temperature, *Revista Construcțiilor și Materialelor de Construcție*, X, No. 12, 1958, pp. 619–620.

5.19. STAICU, C. I., On the formula used for the determination of machine part wear, *Metalurgia și Construcția de Mașini*, XI, No. 3, 1959, pp. 196–198.

5.20. STAICU, C. I., Thermic wave damping, Metalurgia și Construcția de Mașini, XI No. 11, 1959, pp. 970–972.

5.21. STAICU, C. I., New relations for the study of fluid atomizing, *Energetica*, XI, No. 7, 1963, pp. 292–295.

5.22. STAICU, C. I., Diameter of drain fittings of water pipes, *Energetica*, XI, No. 12, 1963, pp. 628–630.

5.23. STAICU, C. I., A mathematical model for fluid boiling, *Buletinul Institutului de Studii și Proiectări Energetice*, X, No. 2, 1967, pp. 125–132.

5.24. STAICU, C. I., Diameter of the water distributing pipe in the cooling tower, *Energetica*, XVI, No. 11, 1969, pp. 510–512.

5.25. STAICU, C. I., Etude du puits en écoulement non permanent par l'analyse dimensionnelle générale, *La Houille Blanche*, XXIV, No. 6, 1969, pp. 656.

5.26. STAICU, C. I., Determination, through dimensional analysis, of steam consumption of the condenser ejector, *Energetica*, XVIII, No. 12, 1970, pp. 538–540.

5.27. STAICU, C. I., Nouvelle relation pour l'écoulement non permanent dans les puits, *Revue Roumaine des Sciences Techniques-Mécanique Appliquée*, XVI, No. 1, 1971, pp. 81–90.

5.28. STAICU, C. I., Pressure drop through the pipes, *Energetica*, XX, No. 9, 1972, pp. 422–423.

5.29. STAICU, C. I., Development of dimensional analysis with application to the study of heat convection – a doctorate thesis, 1972.

5.30. STAICU, C. I., Note sur la convection de la chaleur dans un milieu poreux, *Revue Générale de Thermique*, XVI, No. 185, 1977, pp. 453–459.

5.31. STAICU, C. I., Linear pressure drop in laminar flow, *Energetica*, XXVI, No. 2, 1978, pp. 52–54.

5.32. ŞTEFĂNESCU, D. and GRÜNWALD, B., *Heat Transfer and Gas Dynamics*, Bucharest, Editura Didactică şi Pedagogică, 1965.

5.33. TCHERNOV, A. and BESSREBRENNIKOV, N., *Les Fondements de la Thermique et de l'Hydraulique*, Moscow, MIR, 1971.

5.34. UNGUREANU, C., Fuel atomizing by means of low-pressure pneumatic injectors. *Energetica*, X, No. 8, 1962, pp. 313–319.

5.35. Hütte-Handbook of the Engineer, Bucharest, Editura AGIR, 1947.

INDEX